Benjamin Butz

Yttria-doped zirconia as solid electrolyte for fuel-cell applications

Benjamin Butz

Yttria-doped zirconia as solid electrolyte for fuel-cell applications

Fundamental aspects

Südwestdeutscher Verlag für Hochschulschriften

Impressum/Imprint (nur für Deutschland/ only for Germany)
Bibliografische Information der Deutschen Nationalbibliothek: Die Deutsche Nationalbibliothek verzeichnet diese Publikation in der Deutschen Nationalbibliografie; detaillierte bibliografische Daten sind im Internet über http://dnb.d-nb.de abrufbar.

Alle in diesem Buch genannten Marken und Produktnamen unterliegen warenzeichen-, marken- oder patentrechtlichem Schutz bzw. sind Warenzeichen oder eingetragene Warenzeichen der jeweiligen Inhaber. Die Wiedergabe von Marken, Produktnamen, Gebrauchsnamen, Handelsnamen, Warenbezeichnungen u.s.w. in diesem Werk berechtigt auch ohne besondere Kennzeichnung nicht zu der Annahme, dass solche Namen im Sinne der Warenzeichen- und Markenschutzgesetzgebung als frei zu betrachten wären und daher von jedermann benutzt werden dürften.

Verlag: Südwestdeutscher Verlag für Hochschulschriften GmbH & Co. KG
Dudweiler Landstr. 99, 66123 Saarbrücken, Deutschland
Telefon +49 681 37 20 271-1, Telefax +49 681 37 20 271-0
Email: info@svh-verlag.de
Zugl.: Karlsruhe, Karlsruher Institut für Technologie (KIT), Diss., 2009

Herstellung in Deutschland:
Schaltungsdienst Lange o.H.G., Berlin
Books on Demand GmbH, Norderstedt
Reha GmbH, Saarbrücken
Amazon Distribution GmbH, Leipzig
ISBN: 978-3-8381-1775-1

Imprint (only for USA, GB)
Bibliographic information published by the Deutsche Nationalbibliothek: The Deutsche Nationalbibliothek lists this publication in the Deutsche Nationalbibliografie; detailed bibliographic data are available in the Internet at http://dnb.d-nb.de.

Any brand names and product names mentioned in this book are subject to trademark, brand or patent protection and are trademarks or registered trademarks of their respective holders. The use of brand names, product names, common names, trade names, product descriptions etc. even without a particular marking in this works is in no way to be construed to mean that such names may be regarded as unrestricted in respect of trademark and brand protection legislation and could thus be used by anyone.

Publisher: Südwestdeutscher Verlag für Hochschulschriften GmbH & Co. KG
Dudweiler Landstr. 99, 66123 Saarbrücken, Germany
Phone +49 681 37 20 271-1, Fax +49 681 37 20 271-0
Email: info@svh-verlag.de

Printed in the U.S.A.
Printed in the U.K. by (see last page)
ISBN: 978-3-8381-1775-1

Copyright © 2011 by the author and Südwestdeutscher Verlag für Hochschulschriften GmbH & Co. KG and licensors
All rights reserved. Saarbrücken 2011

Contents

1	**Introduction**		**1**
2	**Fundamentals**		**5**
	2.1	Properties of Yttria-Doped Zirconia	5
		2.1.1 $Zr_{0.83}Y_{0.17}O_{2-\delta}$ as Solid Electrolyte Material	5
		2.1.2 Phase Diagram of the Y_2O_3–ZrO_2 System	12
		2.1.3 Electrical Properties of Studied Specimens	24
	2.2	Experimental Techniques	29
		2.2.1 Equipment	29
		2.2.2 Quantitative EDXS	30
		2.2.3 Quantitative EELS	32
3	**Experimental results**		**41**
	3.1	Specimens	41
		3.1.1 Microcrystalline 8YDZ	41
		3.1.2 YDZ Thin Films	42
		3.1.3 Y_2O_3/ZrO_2 Reference Nanoparticles	43
		3.1.4 Reference Specimens for Raman Spectroscopy	46
		3.1.5 Sample Preparation for TEM	48
	3.2	Microcrystalline 8YDZ	49
		3.2.1 Crystal Structure	49
		3.2.2 Phase Distributions	50
		3.2.3 Volume Fraction of Tetragonal YDZ	55
		3.2.4 Chemistry on the Nanoscale	60
		3.2.5 Raman Spectroscopy	83
	3.3	Nanocrystalline YDZ	85
		3.3.1 Microstructure	85
		3.3.2 Structure	94
		3.3.3 Chemistry	99

4 Discussion 103
4.1 Microstructural Evolution and Decomposition of 8YDZ 103
4.1.1 As-Sintered 8YDZ 105
4.1.2 Decomposition of 8YDZ 108
4.1.3 Correlation of Chemical Decomposition and Decrease of Ionic Conductivity 112
4.2 Nanocrystalline YDZ 115
4.2.1 Microstructure 117
4.2.2 Chemistry 119
4.2.3 Crystal Structure 119
4.2.4 Decomposition 120
4.3 Formation and Stability of Phases near c+t — c Phase Boundary . . 121

Summary 125

Acronyms

AC	alternating current
ADF	annular dark-field
A_s	temperature of t ← m transformation on heating
CCD	charge-coupled device
CVD	chemical vapour deposition
DC	direct current
EDXS	energy-dispersive X-ray spectroscopy
EELS	electron energy-loss spectroscopy
FIB	focused ion beam
HAADF	high-angle annular dark-field
HRTEM	high-resolution transmission electron microscopy
MLLS	multi-linear least squares
M_s	temperature of martensitic t → m transformation on cooling
PLD	pulsed laser deposition
PSZ	partially stabilized zirconia
RTA	rapid thermal annealing
SA	selected area
SAED	selected-area electron diffraction
SEM	scanning electron microscopy
SOFC	solid oxide fuel cell
SRO	short-range ordering
STEM	scanning transmission electron microscopy
TEM	transmission electron microscopy
TNP	tetragonal nanoscaled precipitate

XRD	X-ray diffraction
YDZ	Y_2O_3-doped ZrO_2
8YDZ	8.5 mol% Y_2O_3-doped ZrO_2
10YDZ	10 mol% Y_2O_3-doped ZrO_2
ZA	zone axis
c-YDZ	cubic equilibrium YDZ phase
t-YDZ	tetragonal equilibrium YDZ phase
t'-YDZ	metastable tetragonal YDZ phase
t"-YDZ	metastable tetragonal YDZ phase
m-YDZ	monoclinic equilibrium YDZ phase
YSZ	Y_2O_3-stabilized ZrO_2, indicates the stabilization of YDZ in the cubic phase at room temperature
c-ZrO_2	cubic form of ZrO_2
c'-ZrO_2	metastable tetragonal YDZ phase, old name for t"-YDZ
t-ZrO_2	tetragonal form of ZrO_2
m-ZrO_2	monoclinic form of ZrO_2

Chapter 1

Introduction

Working Principle of Solid Oxide Fuel Cells

The function of fuel cells can be defined as a galvanic cell, in which the reagents and the products of the redox reaction are continuously supplied and removed [1,2]. Thus, chemical energy can be directly converted into electrical energy at high efficiency. A simplified scheme of the function of a solid oxide fuel cell (SOFC) is drawn

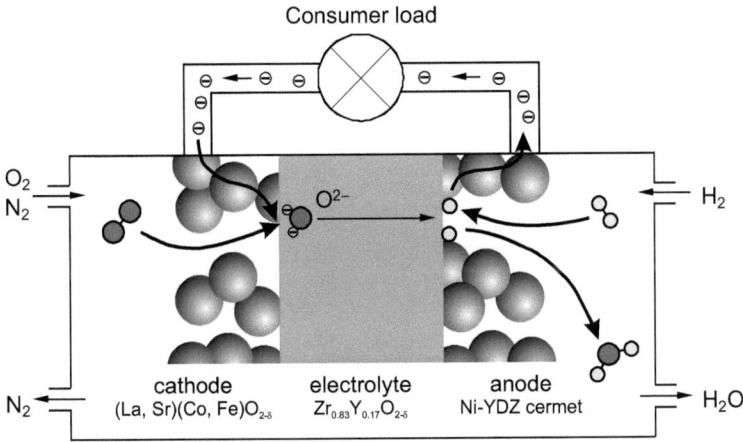

Figure 1.1: Functional scheme of a SOFC

in Fig. 1.1. The reagents, air at the cathode and fuel (H_2 or hydrocarbons) at the anodic side of the cell, are separated by an O-ion conducting ceramic membrane, the solid electrolyte, e.g. Y_2O_3-doped ZrO_2 (YDZ), in order to split the redox reaction into the part reactions, i.e., the reduction of oxygen (taking up of electrons) and the oxidization of hydrogen (release of electron). The chemical potential difference of the oxygen partial pressure between both electrodes initiates the driving force for oxygen compensation through the O^{2-}-conducting electrolyte.

The porous mixed ionic-electronic conducting cathode, e.g. $(La, Sr)(Co, Fe)O_{3-\delta}$, facilitates the dissociation and reduction of oxygen as well as the incorporation of oxygen ions into the electrolyte. The oxidization of fuel takes place close to the triple-phase boundaries at the electrolyte-anode interface, where the O^{2-}-conducting phase (electrolyte grains) meets the electron conducting component of the anode, e.g. Ni, as well as the gas phase containing the fuel and the redox products. Hence, the length of these active sites determines the electrochemical performance of the anode. Due to the exclusive ionic conductivity of the electrolyte the electrons, which are carried from the cathode to the anode by the oxygen ions, are compelled to flow through an outer circuit to perform electrical work.

This work is focussed on the properties of the O^{2-}-conducting electrolyte, i.e. YDZ, as part of the fuel-cell system.

Y_2O_3-doped ZrO_2 as Electrolyte Material

Although many metal-oxide and rare-earth oxide compositions have been examined as ionic conductors during the past decades, zirconia-based materials are still the most common electrolytes at present. The reasons are their beneficial mechanical and electrical properties for SOFC applications and the allegedly well-understood material properties of polycrystalline Y_2O_3-doped zirconia.

The system Y_2O_3–ZrO_2 has been extensively studied for more than 50 years [3–6]. Most of the early publications deal with YDZ as structural ceramic material at various Y contents (mainly below $7\,\text{mol}\%$ Y_2O_3), as periodically reviewed by Garvie [7], Heuer [5] and Rühle [6]. Nevertheless, oxygen-ion conductivity of YDZ was already observed in 15 at% Y_2O_3-doped ZrO_2 by W. Nernst in 1899 [8].

Due to its outstanding mechanical properties depending on the yttrium content and thus the microstructure in combination with the conductivity for oxygen ions, $8.5\,\text{mol}\%$ Y_2O_3-doped ZrO_2 (8YDZ) has been established as one of the most commonly used electrolyte materials for SOFC applications for many years.

Degradation Phenomenon

However, it is a well-known phenomenon that the ionic conductivity of 8YDZ decreases significantly within less than a few 1000 hours at high operating temperatures of 800–1000 °C [9–15]. This degradation as one reason has impeded the commercialization of SOFCs on a big scale until today. However, numerous experimental and theoretical studies have not yet completely clarified the degradation process. Various causes for the degradation of YDZ electrolytes in SOFCs have been proposed in the literature. This will be outlined in detail in § 2.1.1. Several studies have particularly been focussing on the effect of microstructural decomposition of 8YDZ. But experimental evidence of the chemical decomposition has not been given yet. To contribute to the understanding of the decrease of ionic conductivity is one of the aims of this work. Furthermore, the phase diagram of YDZ is still under discussion. This is mainly due to the complexity caused by the numerous stable and metastable phases of YDZ. From the experimental results additional information on boundaries for the chemical and microstructural stability of YDZ as solid electrolyte is derived.

Potential of Nanocrystalline Thin Films

The decrease in physical dimensions down to the nanometre scale is often linked with a dramatic change of the physical and electrochemical properties of materials. Enhanced ionic conduction in nanocrystalline films, if it occurs, is attributed to the grain boundaries. Grain boundaries are regarded as possessing high defect densities and/or enhanced mobilities. Moreover, the adjoining space-charge regions may hold increased ionic defect densities. Positive space charges may lead to an increased oxygen-vacancy density at the grain boundaries. These effects offer a potential means for increasing the ionic conductivity particularly in nanostructured materials with a very high fractional volume represented by the grain boundaries. Nanostructured thin-film electrolytes in anode-supported cells may therefore be able to reduce the portion of ohmic losses significantly.

The use of nanostructured ionic and mixed ionic-electronic conducting materials within fuel cells may facilitate lower operating temperatures and thus enhanced long-term stability of the cells. However, thorough studies of the thin-film processing, the structural and chemical stability of the thin films as well as the transport characteristics and electrochemical properties are necessary to gain an understanding, which is prerequisite for the utilization of nanoionics effects.

The advantages of nanocrystalline YDZ electrolytes, i.e., the enhanced ionic conductivity of YDZ on the nanoscale, is still controversially discussed in literature as outlined in detail in § 2.1.1. Hence, the second part of this work is to study the properties of YDZ on the nanoscale. Therefore nano- and microcrystalline YDZ thin films have been characterized with particular emphasis on the aspects that are relevant for the reliable interpretation of the electrical results.

Goal of this Work and Outline

As mentioned above the degradation of ionic conductivity of common 8.5 mol% YDZ electrolytes is still an obstacle. Several reasons are proposed to cause this degradation of the material. One is the microstructural as well as chemical decomposition of the material as outlined in §2.1.1. Emphasising on the microstructural decomposition, the goal of this work is to visualize the compositional inhomogeneities in 8YDZ after the operation at high temperature by analytical as well as imaging TEM techniques on the nanoscale. This would contributed to the clarification of the decrease of ionic conductivity of 8YDZ.

For the development of SOFC systems in the intermediate-temperature regime of $500\,°C < T < 750\,°C$, nanoscaled thin films are of substantial interest. The second goal of the present work is the study of the role of interfaces, grain size and porosity in nanoscaled electrolyte. Thereto, thorough investigation of the structural, microstructural as well as chemical properties of the thin films are performed to derive an understanding of their charge-carrier transport. Scanning electron microscopy as well as analytical transmission electron microscopy have been utilized to study all relevant aspects of the prepared thin films. Since these thin films with varying mean grain size were prepared by an chemical deposition technique, the preparation only required heat treatments at intermediate temperatures in contrast to sintering of green bodies as used for the preparation of thick-film electrolytes. Hence, this series of specimens is expected to give new insight into the YDZ phase diagram at intermediate temperatures.

Chapter 2
Fundamentals

The first part of this chapter comprises the electrical (§ 2.1.1) as well as the structural (§ 2.1.2) properties of Y_2O_3-doped ZrO_2 (YDZ) with emphasis on the application as solid-electrolyte material. After a short summary of the theory of diffusion in solid electrolytes, i.e., ionic conduction, the investigated stoichiometry of $Zr_{0.83}Y_{0.17}O_{1.915}$, i.e. 8.5 mol% Y_2O_3-doped ZrO_2 (8YDZ), is motivated. The following two open questions related to the use of 8YDZ in fuel cells will be outlined. The obstacle of the degradation of the ionic conductivity of 8YDZ during operation at high temperatures on the time scale of several hundreds to thousands of hours is still not satisfyingly clarified. The second issue is the ionic conductivity of nanocrystalline YDZ, which is proposed to have improved electrical properties in comparion to microcrystalline 8YDZ.

The amount of Y in YDZ does not only influence the transport properties of the material. Moreover, it governs the crystal structure and thus the final microstructure of the YDZ material. Therefore, the second part of this chapter deals with the complicated Zr-rich side of the Y_2O_3-ZrO_2 phase diagram. The present discrepancies will be outlined in detail since these are important for the discussion of the experimental findings of this study. Special emphasis is put on the various phases which may occur in the targeted range of Y contents. The literature overview given is not intended to be exhaustive, since numerous publications have been published on the different issues. The important ones are discussed selectively.

2.1 Properties of Yttria-Doped Zirconia

2.1.1 $Zr_{0.83}Y_{0.17}O_{2-\delta}$ as Solid Electrolyte Material

Yttria-doped zirconia materials are well suited as electrolytes due to their high ionic conductivity for state-of-the-art electrode-supported SOFCs. Moreover, the negligible electronic conductivity even under reducing atmosphere, the electrochemical stability as well as the mechanical properties of YDZ facilitate the use in fuel-cell applications.

W. Nernst already demonstrated high oxygen-ion conductivity in zirconia–magnesia–silica solid solutions in the 1890s [8] under AC and DC conditions, using platinum-wire electrodes. As discussed in his early publication from 1899 [8], he clearly ascribed the electrical conductivity of his specimens at high temperatures to the transport of oxygen ions through the ceramic material in oxygen-containing or wet environments. In the steady-state DC reaction in air he observed a slight but stable reduction of the ceramic solid solutions on the cathodic side, while this phenomenon was not observed in AC experiments. From these observations he concluded a loop of oxygen-ion transport including the incorporation of oxygen on the cathodic and the release of oxygen ions on the anodic electrode. Furthermore, W. Nernst discussed that the conductivity of mixtures of the used materials was much higher in comparison to the conductivity of the pure materials. This obviously has to be attributed to the presence of extrinsic defects in the material mixtures in addition to intrinsic ones.

In the following theoretical discourse about ionic conductivity in such materials, the aspects essential for the discussion are briefly outlined. For a more detailed introduction, the interested reader is referred to the related PhD thesis of C. Peters [16]. The Kröger–Vink notation for lattice elements and point defects in crystal structures is used [17] in order to describe the atomistic reactions related to the replacement of Zr by Y.

Ionic Conductivity

According to current knowledge, ionic transport (oxygen ions) in zirconia-based electrolytes is facilitated by extrinsic vacancies on the anionic sublattice. These are induced by doping of ZrO_2, which means by the replacement of Zr^{4+} by cations (M=Y, Sc, Ca, Mg and others) with lower oxidation states (M^{3+}, M^{2+}) but suitable ionic radii. This conclusion of a completely filled cationic but a deficient anionic sublattice was drawn early by F. Hund in 1951 [18].

For this work, the doping with Y^{3+} ions is essential since Y_2O_3-doped ZrO_2 is still, in addition to Sc-codoped zirconia, one of the most commonly used electrolyte materials for high- and intermediate-temperature SOFC applications.

$$Y_2O_3 \longrightarrow 2Y'_{Zr} + 3O^x_O + V^{\bullet\bullet}_O \tag{2.1}$$

The oxygen vacancies then arise from charge neutrality of the crystal according to Eq. 2.1. In this equation, Y'_{Zr} represents the Y ions on regular cationic sites (instead of Zr^{4+} ions) with a relative charge of -1. O^x_O stands for oxygen ions on regular oxygen sites and $V_O^{\bullet\bullet}$ represents the double positively charged (with respect to the charge of regular oxygen ions) oxygen vacancies. Thus, two dopant Y ions are necessary to introduce a single oxygen vacancy. The concentration of vacancies on the anionic sublattice can be calculated from the concentration of Y ions on the cationic one and arises in the investigated dopant range to a few percent (Eq. 2.2).

$$[V_O^{\bullet\bullet}] = \frac{1}{2}[Y'_{Zr}] \qquad (2.2)$$

The concentration of thermally induced intrinsic defects in contrast is negligible in the targeted temperature range, which explains the low conductivity of pure zirconia. The conductivity as a product of charge-carrier concentration and mobility not only depends on the total concentration of the oxygen vacancies $[V_O^{\bullet\bullet}]$. Furthermore, it depends on the mobility of the oxygen ions or rather of the oxygen vacancies on the anionic sublattice as well as along the grain boundaries (in polycrystalline material). With the assumption of a constant mobility of charge carriers, one would expect an linear increase of ionic conductivity with increasing Y content. However, in contrast to this expectation, the system exhibits a maximum of ionic conductivity followed by the decrease of ionic conductivity with further increasing Y doping as shown in the following.

The mobility, however, is determined by the affinity of the dopant cation to form clusters, by the ion radius of the cation with respect to the host lattice, and by the microstructure of the compound consisting of grains, internal interfaces (e.g. grain boundaries, pores, cracks).

Maximum of Ionic Conductivity of YDZ

Early systematic studies on the DC conductivity of polycrystalline YDZ with respect to Y content and temperature were performed by Dixon et al. [20] and Strickler & Carlson [19]. Both groups found a maximum of conductivity for YDZ at 1000 °C ($0.16\,\mathrm{Scm^{-1}}$ [20], $0.10\,\mathrm{Scm^{-1}}$ [19]) at about 8–9 mol% Y_2O_3, which was confirmed by Casselton [21] ($0.13\,\mathrm{Scm^{-1}}$) and Gibson, Dransfield & Irvine [22] ($0.16\,\mathrm{Scm^{-1}}$) for the same temperature. The experimental data presented in the publications by Dixon et al. [20], Strickler & Carlson [19], and Casselton [21] is summarized in Fig. 2.1, where the maximum of conductivity can be recognized at this dopant content nearly independent on temperature. This is close to the low-yttria boundary of the cubic YDZ field at lower temperatures as will be discussed later (cf. § 2.1.2, page 23).

Hence, 8 mol% Y_2O_3-doped ZrO_2, which was considered to be pure cubic at that time, was established as compromise electrolyte material with sufficiently high (regarding the *cubic* stabilization) but as low as possible (maximum of ionic conducitivity) Y content in the 1980s. In contrast, Gibson et al. [22] found the maximum of ionic

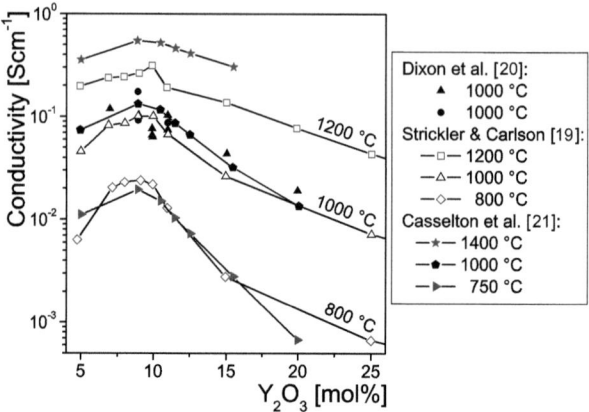

Figure 2.1: Conductivity of YDZ depending on Y_2O_3 content and temperature, redrawn after Strickler & Carlson [19].

conductivity at 7 mol% Y_2O_3.
While the increase of conductivity below 8.5 mol% can be understood by the increase of oxygen vacancies induced by the Y ions, the formation of various types of clusters (also called complexes, associates, defects) between $V_O^{\bullet\bullet}$ and the Y^{3+} and Zr^{4+} ions is discussed as a reason for the decrease of of the initial conductivity of YDZ with higher Y contents, as depicted in Fig. 2.1 [23–28].

Degradation of Ionic Conductivity of 8YDZ Electrolytes

It has turned out early during long-term stability experiments that the conductivity of 8–9 mol% YDZ significantly decreases within few thousand hours during operation at temperatures of about 1000 °C. Indications for the degradation were found by Strickler & Carlson [19] and Baukal [29] in *cubic* stabilized 9 mol% YDZ. In these early studies, the reasons for the decrease are only vaguely discussed, since experimental data on microstructure and chemistry are sparse. Impedance spectroscopy was not yet utilized by the investigators in order to separate grain and grain-boundary contributions to the overall conductivity of the studied materials.

State of knowledge is a decrease of ionic conductivity of 8.5 mol% YDZ by about 40 % after 2500 h at 950 °C [15]. For 8 mol% YDZ, a decrease of 40 % has been reported after 1000 h at 1000 °C [9,13,14]. Comprehensive investigations on the reasons for the degradation phenomenon has been performed by several groups (Ciacchi et al. [30,31], Kondoh et al. [9,32–34], Appel et al. [10], Balakrishnan et al. [11], Hattori et al. [12,13], Haering et al. [14], and Butz et al. [15,35]).

As will be outlined in the following section, various microstructural and chemical changes in YDZ have been discussed and proposed to induce the decrease during operation. However, experimental evidence for most of them is relatively sparse or missing. Second, misinterpretations of experimental findings as well as inappropriate choice of experimental techniques have led to inconsistencies regarding the explanation of the degradation phenomenon. While in the early studies on the degradation, impurity segregation may have had played a major role, diverse reasons have been discussed in recent decades as will be outlined later.

From impedance spectroscopy it is known that the degradation of highly pure 8YDZ has to be attributed to an bulk mechanisms. This technique allows to distinguish between bulk and grain-boundary contributions to the overall impedance. Several groups clearly have demonstrated an increase of the bulk impedance, while the grain-boundary contribution has been found to remain unchanged after ageing at about 1000 °C [11, 15, 31].

Various possible causes for degradation of the bulk conductivity of 8YDZ were suggested. From the theoretical point of view, both the decrease of the concentration of mobile minority charge carriers and/or their mobility must cause the decrease of conductivity during heat treatment. Thus, different explanations are deduced from experimental findings by the groups who have dealt with this problem:

- Phase separation/decomposition:
  As the first one, Baukal discussed the possible decomposition and phase separation of 9 mol% YDZ in the c+t two-phase field into less-conducting c-YDZ and t-YDZ (cf. §2.1.1) in 1969 [29]. But from XRD powder diffractograms he excluded the phase separation of his investigated specimens, since he had not observed additional reflections due to non-fluorite type phases. Ciacchi et al. took this question about phase separation up again in 1991 [30, 31], when a lot of microstructural studies on phase transformations and the resulting microstructures in the system were performed by TEM in the 1980s (cf. §2.1.2, page 17). They identified the precipitation of t-YDZ during heat treatment at 1000 °C for 2000 h as the main reason for degradation of ionic conductivity in 8 mol% YDZ. The distribution of the phase was visualized by dark-field TEM imaging using intermediate magnifications. The authors stated the size of t-YDZ after the heat treatment to be about 6 nm. Neither any analytical evidence for the decomposition of 8 mol% YDZ accompanying the phase separation nor any change of the lattice parameters of the tetragonal regions with respect to the cubic phase, which has to be expected for the c $\rightarrow$ c+t transformation (cf. §2.1.2, page 17), is given in the manuscript [30].

- Oxygen-ion displacement, formation of t"-YDZ:
  Similar tetragonal-type regions were found by Balakrishnan et al. [11] after an-

nealing 8 mol% YDZ at 1000 °C for 1000 h. From dark-field imaging and electron diffraction, they concluded the formation of 10–30 nm regions, which they interpreted in terms of the precipitation of t"-YDZ. Since no evidence for the formation of this phase was found by XRD, they utilized Raman spectroscopy to detect the appearance of tetragonal-type Raman modes during annealing.

They speculated that these regions with displaced oxygen-ion sites in t"-YDZ may have a negative impact on the mobility of oxygen vacancies, leading to the degradation of ionic conductivity of the material. But no information is given on the identification of the t"-YDZ phase. The estimated size of the tetragonal regions from the presented dark-field image appears doubtful because the TEM sample thickness in the investigated regions seems to be too thick to separate single regions. Signal averaging due to the intermediate TEM sample thickness may have had caused an overestimation of the size of the coarsened regions because overlapping regions cannot be distinguished in the projected image intensity. Hattori et al. [12, 13] came to similar conclusions regarding the formation of t"-YDZ.

- Short-range ordering:
  Local changes of the crystal structure on the scale of a few atoms, named short-range ordering (SRO), were stated to cause the decrease of conductivity of 8 mol% YDZ by Kondoh et al. [9, 32–34]. They proposed the local clustering of oxygen vacancies during heat treating YDZ, as it has been found in highly doped YDZ. Such clustering would result in the local trapping of vacancies. This means that these vacancies with increased binding energies may not contribute any more to ionic conduction of the material [14, 32, 33].

In summary, most of the proposed explanations for the degradation are still controversially disputed since experimental proof for them is lacking. Hence, this problem is not yet clarified completely. In order to demonstrate the chemical decomposition of heat-treated (after the degradation) 8YDZ in the c+t two-phase field analytical, TEM on the nanoscale was utilized.

Reduction of Particle Size: *Nanoionics*

The second issue of this study, closely related to the above-mentioned one, is the grain-size dependency of the ionic conductivity of 8YDZ. In general, Tuller [36, 37] and Maier [38–43] pointed out that a substantial reduction of the mean grain size of ionic conducting materials to the nanoscale may lead to an enhanced ionic conductivity through grain-boundary charging effects.

Answering this question, one has to distinguish between two different types of size effects [39]. Trivial effects, which equally occur in microcrystalline as well as nanocrystalline materials, are augmented by the high interfacial density, i.e. grain-boundary density, in nanocrystalline materials. The modification of grain-boundary properties on the nanoscale by varying segregation of impurities and/or dopant cations to the grain boundaries also counts as a trivial size effect. As true size effect, the intrinsic modification of the local physical properties of the grain-boundary regions due to the superposition of the space-charge layers of grain boundaries is mentioned by Maier [39]. This would even lead to enhanced conduction perpendicular to the grain boundaries compared to the microcrystalline material. A more detailed overview of the theory of the proposed space-charge model in ionic conductors can be found in various publications by Maier, Fleig, and Tuller.

As indicated in the following part, various chemical as well as physical deposition techniques have been utilized to prepare nanocrystalline YDZ thin films on various types of substrates. The development of partially stabilized YDZ films for the use as thermal barrier coatings (TBC) was intensified by NASA (National Aeronautics and Space Administration, USA) at the end of the 1970s, as outlined by Miller et al. [44]. The activities to prepare cubic stabilized YDZ thin films for electrode supported fuel cells have expanded within the last decade [45–69]. Most of the publications are mainly based on few results obtained from XRD or SEM as common techniques, which can be easily conducted without complicated sample preparation. Combined XRD and SEM analyses to obtain information on film quality, i.e. adhesion and presence of cracks, film thickness, grain size, crystal structure, and degree of crystallinity, have been conducted by few groups [46, 50, 51, 63, 64, 69]. But detailed microstructural, structural, and chemical characterization by TEM is unusual. This may be attributed to the complicated sample preparation as well as the complexity of the techniques. Only Ning et al. [70] and Briois et al. [67] have presented comprehensive TEM results of their 4.5 mol% YDZ and 8 mol% YDZ thin films prepared by plasma spraying and DC magnetron sputtering, respectively.

For Y_2O_3-doped ZrO_2 around 8.5 mol% Y_2O_3, an enhanced conductivity with decreasing mean grain size and thus drastically increased fraction of grain-boundary related volume has been reported by several groups. Kosacki et al. [52] reported for nanocrystalline YDZ, prepared by spin coating using a polymer precursor, an increase of two orders of magnitude in conductivity compared to polycrystalline and single crystalline YDZ. In a later work, Kosacki et al. [57] observed significantly enhanced electrical conductivity of highly textured 10 mol% YSZ thin films (thickness t < 60 nm) deposited onto MgO substrates by PLD. Similar effects of enhanced oxygen diffusion in nanocrystalline 6.9 mol% YDZ specimens were reported by Knöner et al. [55] who found an increase by three orders of magnitude of oxygen diffusion along the grain boundaries of DC cosputtered YDZ compared to the bulk.

Mondal et al. [50] measured enhanced specific grain-boundary conductivity in nanocrystalline tetragonal 1.7 mol% and 2.9 mol% YDZ in comparison to microcrystalline material, which they attributed to grain-size dependent limited segregation of blocking impurities in the nanocrystalline specimens [47, 50]. Similar findings regarding minimized segregation of blocking impurities are published by Kosacki et al. [49].

Compared to bulk 8YDZ, Park et al. [64] attributed the lower conductivity of 8YSZ thin films, prepared by magnetron cosputtering of Zr and Y, to poor crystallinity and enhanced grain-boundary blocking. These explanations for the decreased conductivity are only mentioned in the manuscript. Similar results were published by Xia et al. [51], who found very low conductivity of gas-tight 8YDZ thin films prepared by a sol-gel based route. The decrease of ionic conductivity in nanocrystalline 8YDZ with respect to microcrystalline 8YDZ (one order of magnitude) was also observed by Seydel [71], who utilized chemical vapour deposition (CVD) to prepare dense YDZ thin films on Ni/YDZ anode as well as LSM cathode substrates. Ramamoorthy et al. [72] reported the shift of the maximum of the ionic conductivity of nanocrystalline YDZ towards lower dopant contents, i.e., 3 mol%.

In contrast, Joo & Choi [63], Chen et al. [46] reported similar conductivity of their prepared thin films with 7.5 mol% and 8 mol% Y_2O_3, respectively, in comparison to microcrystalline material.

The role of impurity segregation and interfacial contribution is inconsistently observed (cf. §2.1.2, page 22). Some groups have suggested that an intergranular siliceous phase accounts for the very low grain-boundary conductivity (cf. §2.1.2).

In summary, the role of internal interfaces, i.e. grain boundaries, in YDZ is still controversially discussed. A thorough structural, microstructural, and chemical characterization of the investigated specimens, which is missing in most publications, is indispensable in order to interpret the electrical characterization. To reliably determine the specific grain-boundary and bulk conductivity, comprehensive knowledge of not only the grain-boundary density but also porosity, pore-size and grain-size distributions, and the distribution of the present phases is essential. As will be shown, even the crystal structure and local chemistry in the grains can have a strong influence on bulk conductivity.

2.1.2 Phase Diagram of the Y_2O_3–ZrO_2 System

Y_2O_3-doped zirconia is one of the most relevant structural ceramic materials for high-temperature applications, e.g. thermal-barrier coatings, due to its outstanding mechanical and chemical properties. This is best exemplified in the publication of Garvie et al. [7] that carries the pictorial title *Ceramic steel?*. Especially partially stabilized zirconia (PSZ) and completely stabilized zirconia with Y_2O_3 contents in the zirconia-rich region of the phase diagram have been intensively studied for more than a century.

The stabilization of the cubic high-temperature phase at room temperature at Y contents ≥ 8.5 mol% has been stated by many authors. As it has turned out in recent studies on namely cubic stabilized 8YDZ, the addition of 8.5 mol% Y_2O_3, which is used for this study, is not sufficient to completely stabilize YDZ in the cubic structure. For a thorough discussion of the experimental findings of this work, a detailed summary about the known phases of ZrO_2-rich (< 10 mol% Y_2O_3) YDZ, their transformation behaviours and the resulting microstructures is indispenseble. In order to summarize the aspects of the YDZ phase diagram, a schematic drawing of the Zr-rich regions of the Y_2O_3–ZrO_2 phase diagram is given in Fig. 2.2.

Early crystallographic studies of namely pure ZrO_2 are summarized by Ruff & Ebert [74]. Due to the difficulty of obtaining highly pure zirconia at that time, strong discrepancies arose regarding the crystal structure of this material, as outlined in that publication. In a few studies, pure zirconia was found to be of cubic structure at room temperature [75]. But it seems to be more likely that impurities (e.g. calcia, magnesia, and others), which had been assumed to influence the crystal structure, may have had a stabilizing effect on the cubic high-temperature phase as known nowadays.

According to the phase diagram (Fig. 2.2), pure zirconia exhibits three stable crystal structures depending on temperature: At high temperatures, ZrO_2 is cubic (c-ZrO_2, space group $Fm\bar{3}m$), while it is tetragonal (t-ZrO_2, space group $P4_2/nmc$) at intermediate temperatures. At lower temperatures, zirconia transforms into the monoclinic (m-ZrO_2, space group $P2_1/c$) phase.

Pure Y_2O_3 is cubic (Y_2O_3, space group $Ia3$) below 2300 °C. Table 2.1 comprises crystal-structure data of pure Y_2O_3 and ZrO_2 as well as of various phases of YDZ used for simulations and discussion in the present study. Both polymorphs of ZrO_2, which can be found at lower temperatures, m-ZrO_2 and t-ZrO_2 have been described early, as summarized by Ruff & Ebert [74]. The crystal structure of pure m-ZrO_2 was refined by McCullough & Trueblood [76]. Smith & Newkirk [77] and Hann [78] confirmed these lattice parameters with slightly higher precision.

Polymorphism of ZrO_2

The tetragonal form of zirconia (t-ZrO_2) is described as a slightly deformed fluorite-type cell [79]. A first detailed crystallographic study of the tetragonal form of zirconia was presented by Teufer [79], who described the unit cell of the tetragonal form by a distorted oxygen sublattice (shift from the original positions of oxygen ions in the

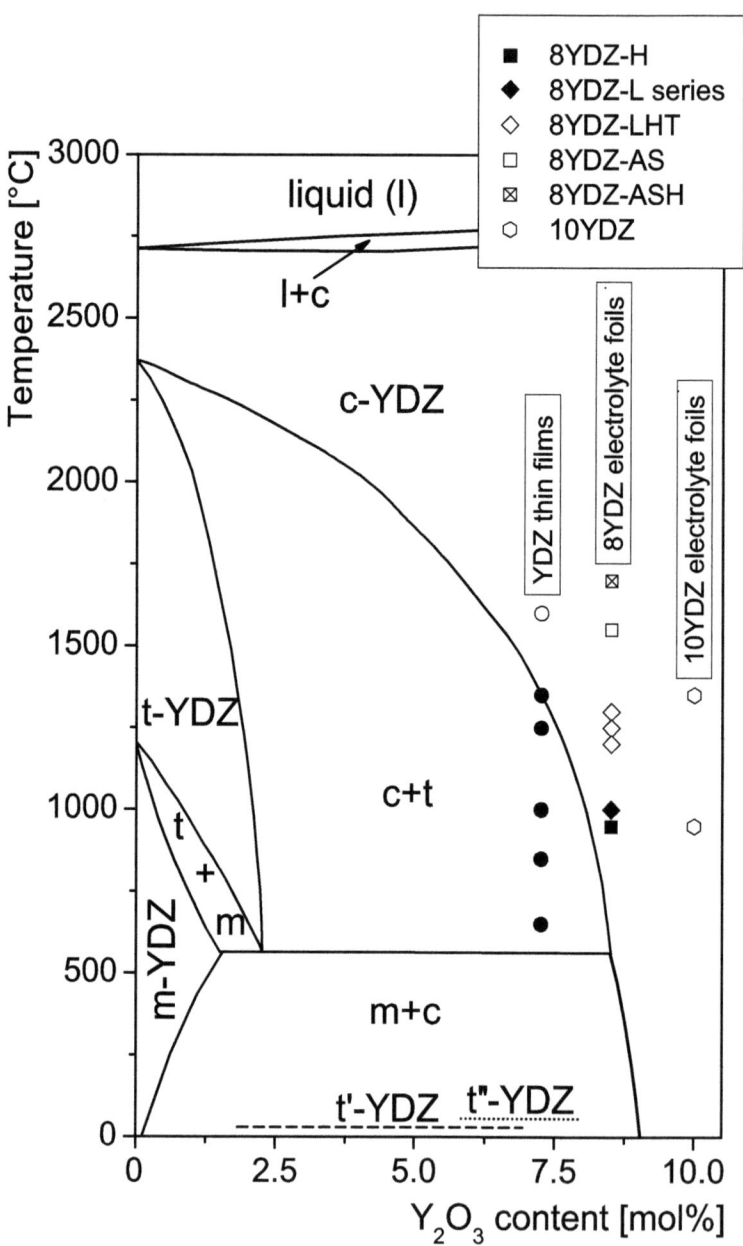

Figure 2.2: Schematic drawing of the Y_2O_3–ZrO_2 phase diagram in the zirconia-rich region, redrawn after Scott [73].

	Lattice parameters [nm]	Space group	Positions of ions	Ref.
m-ZrO$_2$	a_m=0.515 b_m=0.521 c_m=0.532 β=99.23°	P2$_1$/c	Zr^{4+} (0.2754, 0.0395, 0.2083) O^{2-} (0.0700, 0.3317, 0.3447) O^{2-} (0.4496, 0.7569, 0.4792)	[75]
c-Y$_2$O$_3$	a_c=1.0604	Ia3	Y^{3+}(1) (0.2500, 0.2500, 0.2500) Y^{3+}(2) (0.9672, 0.0000, 0.2500) O^{2-} (0.389, 0.154, 0.378)	[76]
c-YDZ (in 8YDZ)	a_c=0.5142	Fm$\bar{3}$m	Zr^{4+}, Y^{3+} (0, 0, 0) O^{2-} (0.25, 0.25, 0.25)	[72, 77]
t'-YDZ	$a_{t'}$, $c_{t'}$ cf. Fig. 2.5	P4$_2$/nmc		[72, 144]
t''-YDZ (in 8YDZ)	$a_{t''}$=0.3636 $c_{t''}$=a_c=0.5142	P4$_2$/nmc	Zr^{4+}, Y^{3+} (0, 0, 0) O^{2-} (0, 0.5, 0.237)	[77]

Table 2.1: Structure data of ZrO$_2$, Y$_2$O$_3$, and YDZ phases.

cubic unit cell along c-direction), while the cationic positions of the pseudo-fluorite cell are unchanged (*Teufer*-cell). This pseudo-fluorite cell is schematically shown in Fig. 2.3, where the shift of the oxygen ions along the c-axis is slightly amplified for better visualization. He reported a ratio $\frac{c_t}{a_t} \approx 1.02$ at 1250 °C [79].

The martensitic transformation (shear-type) t $\leftrightarrow$ m at lower temperatures, which is characterized by a large thermal hysteresis between M$_s$ (*martensitic start temperature*, on cooling) around 1000 °C and A$_s$ (in analogy to the transformation in steel: *austenitic start temperature*, on heating) around 1200 °C [74, 80–84]), is described by several steps of atomic rearrangements within the unit cell [80, 81, 85]. It is accompanied by an increase of the volume of the unit cell (t $\rightarrow$ m) and thus the volume of the transforming individual grains in polycrystalline zirconia. Therefore, it was identified as the reason for rupture of pure ZrO$_2$-made pieces [74]. Depending on heating and cooling rates, the coexistence of t-ZrO$_2$ and m-ZrO$_2$ has been found by several authors, as summarized by Subbarao et al. [86]. A systematic study on the nature of the transformation by TEM including comprehensive crystallographic aspects was presented by Bansal & Heuer [87, 88]. They directly showed the similarities of the transformation in ceramic ZrO$_2$ to the transformation, which is well-known for Fe-based alloys and steel.

This transformation is more sluggish than the c $\leftrightarrow$ t transformation (cf. Fig. 2.4), caused by the developing strain that strongly affects the rate of transforming grains in the material.

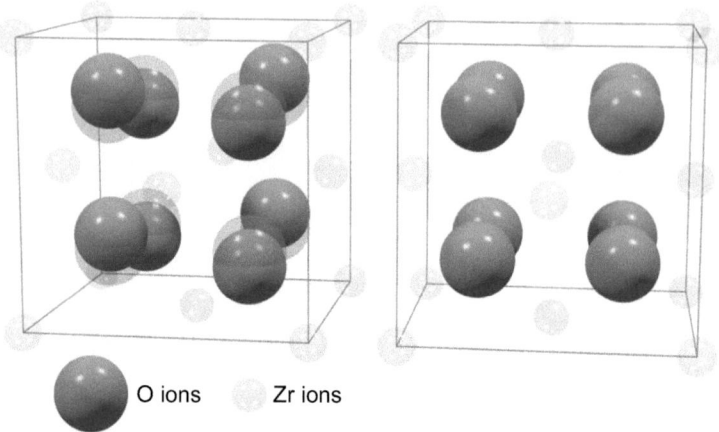

Figure 2.3: Schematic drawing of the tetragonal unit cell (*Teufer*-cell) of ZrO_2 (left) in comparison to the fluorite-type structure (right) of the cubic form of ZrO_2, for better visualization the shift of the oxygen ions is exceedingly shown.

As result, considerable disagreement about the transformation temperature in the range of about 1000–1200 °C exists among the reported values, as summarized by Patil & Subbarao [81] and Yashima and co-workers [89].

The existence of the fluorite-type high-temperature phase was verified by Smith & Cline [90] at temperatures above 2285 °C applying high-temperature XRD. Rouanet [91] observed a slightly higher temperature of 2340 °C for the reversible transformation of c-ZrO_2 into t-ZrO_2 (c ↔ t), which still holds today.

A comprehensive summary of the m ↔ t and t ↔ c transformation temperatures of pure ZrO_2 can be found in the publication on theoretical modelling of the Y_2O_3–ZrO_2 system presented by Chen et al. [92]. Indications for the occurrence of metastable phases even in pure zirconia were reported by Garvie [93] who attributed the preservation of the tetragonal form at lower temperatures in fine-grained ZrO_2 to the small size of the crystallites and thus high interfacial energy, which would accompany the phase transformation to the monoclinic structure. As will be shown, such size-dependent stabilization plays a key role in the explanation of the outstanding mechanical properties of partially stabilized zirconia.

Structure of Y_2O_3

A detailed XRD study of pure Y_2O_3 single crystals by Paton & Maslen [94] has revealed face-centred cubic structure for Y_2O_3. This structure of Y_2O_3 as sesquioxide

(M_2O_3) is similar to the cubic rare earth oxide structure that is known for most rare-earth oxides [95]. The structure was confirmed by following investigations [96].

Miscibility Gap

Since the solvuses of Y ions in t- and m-ZrO_2 are limited to a few mol%, a miscibility gap exists in the zirconia-rich region of the Y_2O_3–ZrO_2 system, as depicted in the phase diagram (Fig. 2.2). Necessarily, two-phase fields (t+m, c+t, c+m) arise, in which the respective phases are expected to coexist in thermodynamic equilibrium with volume fractions according to the lever rule.
With increasing Y content, the temperature of the c $\leftrightarrow$ c+t transformation decreases. If a sufficient amount of Y_2O_3 is added, the cubic high-temperature phase may be stabilized at room temperature, as indicated on the right-hand side of Fig. 2.2. The transformation of t-YDZ into m-YDZ has been reported to appear at lower temperature than in pure zirconia.

Phase Diagrams

Numerous revised phase diagrams of the zirconia-rich region as summarized in Fig. 2.4 have been published from the early 1950s until today, summarizing and discussing the numerous studies on the ternary Y_2O_3–ZrO_2 system [3, 4, 73, 97–105]. Yashima et al. [89] discussed discrepancies, which had arisen, in depth. In addition, they presented a comprehensive discussion about metastable phase transitions on the basis of proposed Gibbs-free-energy functions to explain the occurrence of both metastable phases t'-YDZ and t"-YDZ in this region of the phase diagram.
Both diffusion-controlled transformations c $\rightarrow$ c+t and c $\rightarrow$ c+m have been observed to take place in a sluggish manner, because bulk self-diffusion of Y and Zr on the cationic sublattice in YDZ is known to be very slow at temperatures below 1300 °C [106–111].

Nucleation and Growth of t-YDZ

The decomposition of c-YDZ during heat treatment in the c+t two-phase field is described by a coherent precipitation mechanism combined with the growth of tetragonal regions in the cubic matrix of the grains [82, 102, 112–116]. The molar volume change during this transformation is nearly zero [113]. Hence, negligible strain appears during the transformation. The resulting modulated morphology [112] is of lenticular t-YDZ precipitates coherently twinned with the Y-enriched cubic matrix,

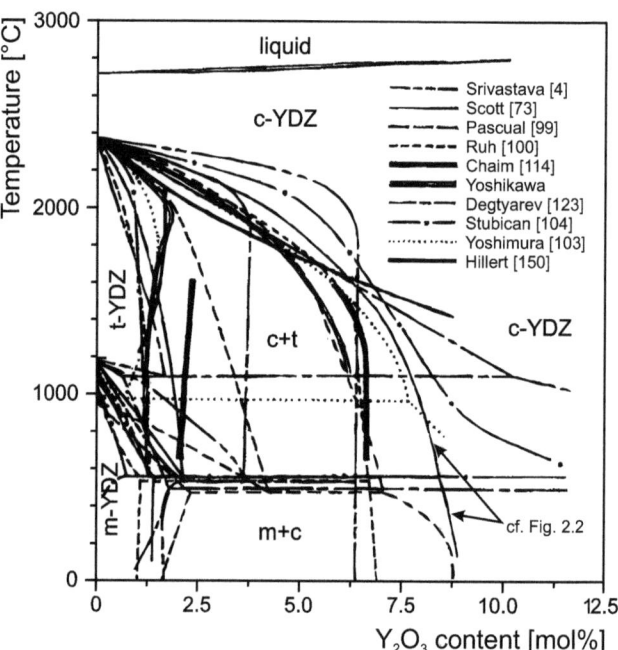

Figure 2.4: Revised phase diagrams of the Y_2O_3–ZrO_2 system, redrawn after Yashima et al. [89].

the twin plan being {110}.
Most of the studies were performed at temperatures in the range of 1200–1600 °C using YDZ with Y-contents in the range of about 4 to 7 mol%. From these experiments performed at temperatures above 1200 °C, the boundaries of the c+t phase field were delimited as summarized by Heuer & Rühle [113].

Spinodal Decomposition

As competitive possibility in achieving thermal equilibrium during heat treatment of c-YDZ in the centre of the c+t two-phase field, spinodal decomposition has been observed by a few groups [117–120]. Direct evidence for the local variation of the Y content by analytical TEM is only given for 6 mol% YDZ by Katamura et al. [119] and Shibata et al. [120] after heat treating their specimens at 1200 °C for 126 h.

The region of the chemical spinodal was predicted in the centre of the c+t two-phase field by Suto et al. [102] and Sakuma et al. [117]. Due to the fact that spinodal decomposition occurs in the compositional range between the two inflection points next to the maximum of the Gibbs free energy function for a given temperature, i.e. $\frac{d^2G}{dX^2} < 0$, it has to be situated between the boundary lines of the two-phase field in the phase diagram, which are described by the solvuses of Y in c-YDZ and t-YDZ. Thus, this field has to be smaller than the c+t two-phase field.

Experimental studies to clearly define this region are relatively sparse, as outlined above. Especially for temperatures below 1200 °C, no investigations have been published. This is attributed to the low cation mobility and thus long time it takes to carry out such experiments.

Martensitic Transformation c+t ↔ c+m

The dependency of the t → m transformation temperature (boundaries t — t+m as well as c+t — c+m in Fig. 2.2) on the Y content was systematically studied by Geller & Yavorsky [121] and Duwez et al. [97] who applied dilatometry, complex thermal analysis, and high-temperature XRD. Both groups found reduced transformation temperatures of $A_s \approx 550$ °C (on heating) and $M_s \approx 450$ °C (on cooling) at about 3 mol% Y_2O_3 in comparison to pure ZrO_2. According to Gibbs' phase rule, this temperature for the transformation was found to be constant between the c+t and c+m two-phase fields. The hysteresis, i.e., the difference between M_s and A_s, is retained in this range of Y contents [97, 98, 121].

Most of the following experimental studies [4, 98–100, 122] confirmed these temperatures for the c+t → c+m transformation in the range of 500–700 °C up to 7.5 mol%

Y_2O_3 [89, 122]. A higher transformation temperature around 1000 °C was theoretically predicted by Degtyarev & Voronin [123, 124] and Yoshimura [103], but has never been shown experimentally.

Depending on crystallite size and cooling rate, the transformation may be suppressed, known as *transformation toughening* in pure and Y-containing ZrO_2, leading to a metastable state with outstanding mechanical properties [82, 125–130], i.e. high crack resistance and strength. This is due to the fact that the nucleation-controlled transformation, suppressed by high strain-field energy in comparison to reaction enthalpy, is induced at the tip of cracks propagation through the material providing favoured sites for the transformation. Since the transformation is accompanied by an increase of volume of the material at the crack tip, any further propagation of the crack is impeded.

Metastable Phase Transition c → t'

Since the decomposition of YDZ during cooling from the pure cubic phase field through the c+t two-phase field is strongly limited by the mobility of the ions, rapid cooling or quenching YDZ from high temperatures prevents the diffusion-controlled reactions. Furthermore, c-YDZ diffusionlessly transforms from the c-phase into a metastable tetragonal form of YDZ (t'-YDZ). Numerous studies on the instability of t'-YDZ during heat treatment in the c+t two-phase field and the resulting microstructures have been performed [105, 112, 131–144]. As indicated previously, t-YDZ shows lattice parameters independent on the Y-content of YDZ in equilibrium due to the solvuses of Y in t- and c-YDZ (given by the phase diagram). Metastable t'-YDZ is characterized by lattice parameters, which depend on the Y content in the YDZ material. Lefèvre [145] found the linear dependencies of the lattice parameters of t'-YDZ, i.e., $a_{t'}$ and $c_{t'}$, on the dopant content in the range of 2–6 mol% Y_2O_3, as depicted in Fig. 2.5 (open symbols). For better comparability with the fluorite cell of c-YDZ, $\sqrt{2}a_{t'}$ is given instead of $a_{t'}$. The dopand range of 6.5–9 mol% Y_2O_3 was in particular studied in detail by Scott [73], who found similar values for the lattice parameters of t'-YDZ. In addition, they observed the coexistence of c-YDZ and t'-YDZ in the dopant range of 6-6.5 mol% Y_2O_3. While $c_{t'}$ decreases, $\sqrt{2}a_{t'}$ increases with increasing Y content. Both values asymptotically converge to the cubic lattice parameter at about 8.5 mol% Y_2O_3. Hence, the tetragonality $c_{t'}/\sqrt{2}a_{t'}$ (inset in Fig. 2.5) decreases from 1.017 (2 mol% Y_2O_3) to zero at about 8.5 mol% Y_2O_3 (extrapolated value in Fig. 2.5). These findings for t'-YDZ have been confirmed with excellent agreement by several authors [102, 137–139, 141] applying XRD measurements. At Y contents higher than 7 mol% [73, 102, 138, 141], the existence of t'-YDZ has never been shown,

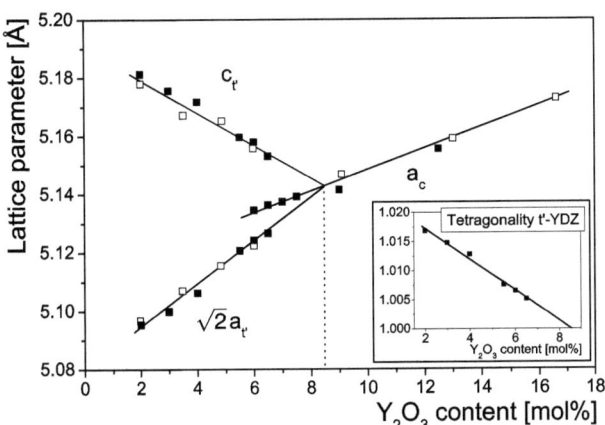

Figure 2.5: Lattice parameters of t'-YDZ ($\sqrt{2}a_{t'}$, $c_{t'}$) and c-YDZ (a_c) vs. Y_2O_3 content (redrawn after Ref. [73, 145]). The tetragonality of t'-YDZ calculated from the given values of Scott [73] is plotted vs. Y_2O_3 content in the inset.

irrespective of the thermal history of the specimens. This might be related to the low sensitivity of XRD, mainly used in these studies, for the detection of small volumes and low volume fractions, if the phase is present at slightly higher Y contents. Despite the measurable differences of the lattice parameters between t'-YDZ and c-YDZ in 6 mol% and 6.5 mol%, as shown in Fig. 2.5, the calculated differences of the volumes of the unit cells for both given dopant contents is less than 1 ‰.

The instability of t'-YDZ during heat treatment in the c+t two-phase field and the consequent diffusional phase separation into c+t-YDZ was described by Scott [73]. Gibson & Irvine [144] estimated a maximum Y content, up to which t'-YDZ can be expected to form. They heat-treated YDZ with various Y contents (3–7 mol% Y_2O_3) at 1500 °C in the c+t two-phase field to facilitate the decomposition. After rapid cooling, they found constant lattice parameters for t'-YDZ independent on the Y content. Furthermore, they observed volume fractions of the final phases according to the lever rule. For that annealing temperature, the estimated solvus of Y in t'-YDZ was ~ 7.5 mol%. Since the curvature of the c ↔ c+t boundary gives the solvus of Y in c-YDZ, this value has to be expected to be higher at lower annealing temperatures (higher solvus of Y in c-YDZ).

Metastable Phase Transition c → t"

A second metastable form of YDZ (t"-YDZ) with tetragonality $c_{t"}/\sqrt{2}a_{t"}=1$ has been found by Zhou et al. [141] in 8 mol% Y_2O_3-doped ZrO_2, who named the observed form c'-ZrO_2. At this Y content, the lattice parameters of the tetragonal form coincide with the lattice parameter of c-YDZ, which also exists at this dopant content. They proposed a distorted oxygen sublattice similar to that of t'-YDZ. More detailed studies on the c ↔ t" diffusionless transformation have been presented by Yashima et al. [146–148]. The proposed crystal structure characterized by a distorted anionic sublattice with shifted oxygen-ion sites was verified by Yashima [147] by means of neutron diffraction. Furthermore, they studied the formation of t"-YDZ at dopant contents in the range of 7–8 mol% by Raman spectroscopy [148]. The coexistence of t'-YDZ and t"-YDZ in the range of 6–8 mol% Y_2O_3 has been described by Yashima, Kakihana, and Yoshimura [89, 147].

Since the tetragonality of t'-YDZ also converges to unity at about 8.5–9 mol% (cf. page 20) as it is characteristic for t"-YDZ, the discussion about the nature of any tetragonal YDZ-phase in this range of Y contents has to be carried out very carefully.

Thermodynamic Modelling

Thermodynamic modelling of the Y_2O_3–ZrO_2 system has been performed by several authors [92, 123, 124, 149–154]. Due to the fact that most of the simulations are based on thermodynamic data mostly obtained at temperatures above 1200 °C [92, 123, 124] strong inconsistencies between the predictions in those publications and experimentally observed phase boundaries exist, especially at temperatures of ≤1000 °C.

Amorphous Grain-Boundary Phases

An important issue for polycrystalline YDZ is the role of amorphous grain-boundary phases, irrespective of the reason (impurity or sintering aid) for their presence, since they have been shown to strongly influence not merely the mechanical, but rather the electronic properties, i.e. ionic conductivity. It is sufficient to say that less or non-conducting phases along grain boundaries will in general impede the transport of oxygen ions perpendicular to the grain boundaries and thus through polycrystalline YDZ.

Since the solubility of Si and Al in the YDZ lattice is limited, glassy silica- and alumina-containing phases are known to segregate at the grain boundaries during the preparation of polycrystalline YDZ. Rühle et al. [155] and Schubert et al. [156] have demonstrated the high solubility of Y in such glassy Si- and Al-containing phases. Therefore, such phases may act as sinks for Y ions. This finally may lead to inhomogeneities in the distribution of Y in the polycrystalline YDZ material [155].

The addition of alumina as sintering aid was found to facilitate faster densification of Ca-doped ZrO_2 [157] and Y-doped ZrO_2 [158], resulting in smaller mean grain sizes of the sintered materials.

Badwal et al. [158] have found a decreasing contribution of the grain boundaries to the overall resistivity of their studied YDZ specimens, which may simply be due to geometrical effects, i.e., a higher grain-boundary density in fine-grained YDZ. No information on the specific resistivity of the grain boundaries depending on the chemical composition and the thickness of the secondary grain-boundary phase is given in the publication.

Depending on the chemical composition of any glassy phase, both, the enhancement as well as the reduction of ionic conductivity in YDZ, are discussed in literature [33, 157–163]. In general, the presence of Si-rich phases at the grain boundaries is accepted to have a negative impact on ionic transport across and along grain boundaries, while the influence of Al on the specific grain-boundary conductivity is much smaller. In order to prevent the complete wetting of glassy SiO at the grain boundaries, Lee et al. [163] presented a high-temperature sintering procedure for YDZ.

The segregation of Y at clean grain boundaries even in high-purity YDZ was studied by Stemmer et al. [161] applying EELS. The authors argue that even despite the absence of segregated Si — this is not simultaneously measured with the local Y and Zr signal in that study — an increase of the Y content at the grain boundaries was observed. No information about the influence of local thickness on the jump ratio the cationic $L_{2,3}$ ionization edges is given in the manuscript. The Y-$L_{2,3}$ intensities of the presented spectra have also not been evaluated quantitatively. Inconsistencies between the presented spectra (different intensity ratios of the white lines of Zr, different slopes of the background in the spectra, different jump ratios) are not discussed in that publication and thus remain unclear.

Stabilization of c-YDZ at Room Temperature

YDZ used as solid electrolyte in high-temperature fuel-cell applications undergoes thermal cycles from room temperature to operation temperature and vice versa. In order to minimize mechanical stresses within the electrolyte itself as well as at the interfaces to the electrodes during heating and cooling, the preferred Y_2O_3-doped ZrO_2 material for such applications does not transform in the targeted temperature range.

Of considerable interest is the minimum Y content to stabilize c-YDZ at low temperatures. Since slow cation diffusion impedes decomposition of YDZ at $T \leq 800\,°C$, the position of the c — c+t phase boundary at this temperature is sufficient to known. The published values range from 5 [18] to 6 [97, 98], 6.5 [4], 7 [141, 144], 7.5 [73], 8 [99–101, 125, 126], 8.5 [164], 9 [122, 148] to 9.2 mol% Y_2O_3 [165]. The strong disagreement about this value is due to several related reasons:

- First of all, it is difficult to achieve thermal equilibrium at temperatures below 1200 °C due to the low diffusivity of cations in YDZ.

- Most of the groups exclusively utilized XRD [4, 18, 73, 97–100, 122, 125, 126, 144, 164], dilatometry [99, 122] and differential thermal analysis [4, 98–100] — methods not sensitive enough to detect small volumes and volume fractions of subsidiary phases — in order to study the crystal structure and phase transformations in their investigated specimens.

- Furthermore, the small scattering factors of additional reflections of t-, t'-, and t"-YDZ in XRD may impede the detection of these phases. This is due to the exclusive shift of the oxygen-ion sites with respect to cubic YDZ. These small scattering factors lead to very weak additional reflections in XRD. Furthermore, the precipitates of the tetragonal-type phase in the specimens studied in this work are of nanoscaled size. This leads to a strong broadening of reflections in XRD. Hence, common X-ray diffractometers are inappropriate to detect small volume fractions and/or nanoscaled regions of tetragonal-type phases in YDZ [11–13].

In summary, it is known that 8.5 mol% YDZ is not completely stabilized at room temperature in the cubic form. A tetragonal-type phase, which is not clearly identified yet, has been described to form during heat treating the material around 1000 °C for several thousands of hours. The three tetragonal forms of YDZ, i.e., t-YDZ, t'-YDZ, and t"-YDZ, can be described as a fluorite-type unit cell with a distorted anionic sublattice as depicted in Fig. 2.3. The shift of the oxygen ions along the c-axis is depending on the nature of the tetragonal-type phase as well as the dopant content. Since the lattice parameters of t'-YDZ converge to the cubic one and thus to the lattice parameters of t"-YDZ at about 8.5 mol%, from the crystallographic point of view it is complicated to distinguish these phases in 8YDZ.

2.1.3 Electrical Properties of Studied Specimens

To contribute to the clarification of the open questions, outlined in the previous section, different types of YDZ specimens have been prepared and characterized for this investigation. The results of the electrical characterization are anticipated here. The preparation of the specimens as well as the applied heat treatments are described in the experimental chapter § 3.1.

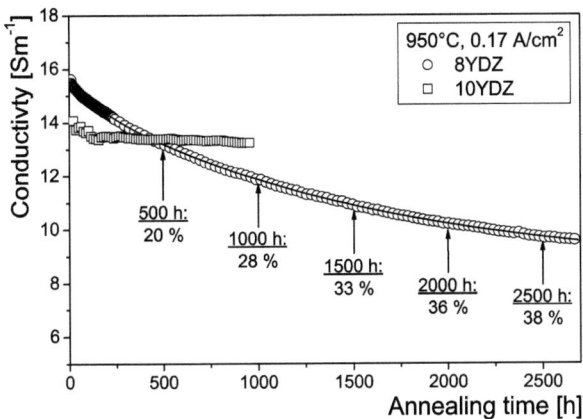

Figure 2.6: Ionic conductivity of 8YDZ and 10YDZ at 950 °C as function of time [15]. The relative reduction of the ionic conductivity of 8YDZ is given for several annealing times.

8YDZ Thick-Film Electrolytes

Thick-film 8.5 mol% (8YDZ) and 10 mol% (10YDZ) Y_2O_3-doped ZrO_2 electrolyte foils as used for electrolyte-supported high-temperature SOFC applications were analyzed in particular to clarify the degradation of ionic conductivity in 8YDZ. Pieces of both types of electrolyte foils were studied with respect to microstructure and chemistry before and after heat treatment, which was applied to emulate the application at commercially relevant conditions, i.e., 950 °C for several thousand hours, while the degradation of conductivity was continuously monitored by means of DC four-probe measurements during the heat treatment.

The evolution of the electrical conductivity of the 8YDZ and 10YDZ specimens was evaluated at the *Institute of Materials for Electrical and Electronic Engineering IWE, University of Karlsruhe* (A. Müller) by four-probe DC measurements during the annealing as described in Ref. [15,166]. Therefore, the substrates were cut into rectangular pieces (10 x 50 mm^2) and contacted by platinum paste and platinum wires. A constant current density was applied via the outer electrodes of the specimens. The conductivity was continuously monitored by the voltage between the inner electrodes as described in Ref. [15] (and publications therein).

Fig. 2.6 shows the overall ionic conductivity of these 8YDZ and 10YDZ specimens as a function of time. While the conductivity of 10YDZ remains nearly unchanged, the considerable decrease of the conductivity of 8YDZ by nearly 40 % within 2500 h at 950 °C can be recognized [15]. To identify the contributions of bulk and grain boundary resistivity on the degradation of the 8YDZ electrolytes, substrate samples were

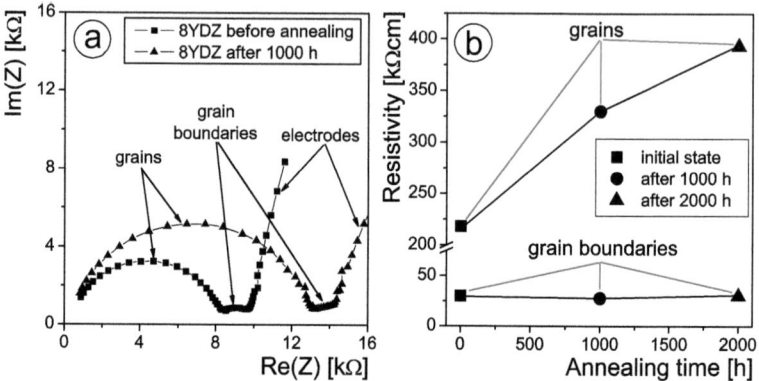

Figure 2.7: a) Impedance spectra of 8YDZ before and after heat treatment at 950 °C for 1000 h, b) grain and grain-boundary contribution to the ionic conductivity [15].

analysed by means of impedance spectroscopy before and after 1000 h and 2000 h of annealing. Therefore, porous platinum electrodes 6 x 6 mm^2 in size were sputtered symmetrically on both sides of the substrates. Impedance spectra were obtained over the frequency range from 10 Hz to 1 Mhz using a Solartron 1260 impedance analyzer. These measurements were conducted in air over the temperature range of 275–450 °C. The impedance spectra, which consist of three semicircles, which were attributed to bulk, grain-boundary, and electrode resistivity, were fitted using the Software MVC-NLS (Multivariate Complex Nonlinear Least Squares). For detailed information, the interested reader is referred to Ref. [15, 166].

Representative impedance spectra of the 8YDZ specimens (as-prepared and after heat treatment at 950 °C for 1000 h), obtained at 300 °C, are given in Fig. 2.7a [15]. The larger arc, attributed to the grains, significantly increases during annealing, while the small arc, which represents the grain boundaries, is almost unchanged after the heat treatment. From these spectra, the grain and grain-boundary contributions to the overall resistivity of the specimens before and after heat treatment are obtained as presented in Fig. 2.7b.

Nanocrystalline YDZ Thin Films

To investigate the influence of grain-size effects on the ionic conductivity of 8YDZ, thin films with varying mean grain sizes between a few nm and the microscale were prepared. The preparation of the thin films, based on a soluble precursor powder, was performed at the *ISC, Fraunhofer-Institut für Silicatforschung, Würzburg*. The electrical properties of the specimens were studied by C. Peters from the *Institute of Materials for Electrical and Electronic Engineering IWE, University of Karlsruhe* in

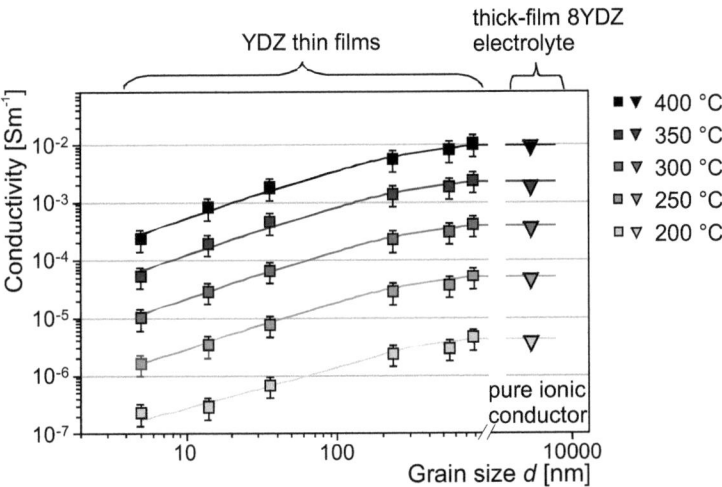

Figure 2.8: Total conductivity of YDZ thin films at temperatures in the range of 200–400 °C depending on mean grain size. The total conductivity of an as-sintered 8YDZ standard electrolyte foil (cf. §2.1.3, first part) is plotted for comparison. The lines are guides to the eye indicating a specific measuring temperature. [16, 167]

the frame of his PhD thesis [16, 167].

A sol-gel based method (cf. §3.1.2) was utilized to prepare thin films of high purity on isolating sapphire substrates. The final dopant content was slightly lower than that of the thick-film electrolytes. This is attributed to the composition of the sol-gel, which was used for the deposition. An annealing procedure, combining a rapid-thermal calcination step with a second heat treatment, was applied to set up the mean grain size and porosity in the final specimens.

The series of specimens was examined regarding crystal structure, microstructure, and chemistry by TEM after the final heat treatments. These results are presented in §3.3. In the following, the results of the electrical characterization of C. Peters [16, 167] are briefly summarized. It was found that the conductivity of the thin films is based on the transport of oxygen ions through the material, i.e., ionic conduction.

The evaluated electrical conductivity of the thin-film samples was found to continuously decrease with decreasing mean grain size. This is visualized in Fig. 2.8 [16], where the total conductivity of the specimens measured at temperatures in the range of 200–400 °C, is plotted against the mean grain size. For comparison, the corresponding values of an as-sintered 8YDZ thick-film electrolyte (cf. §2.1.3, first part) are presented.

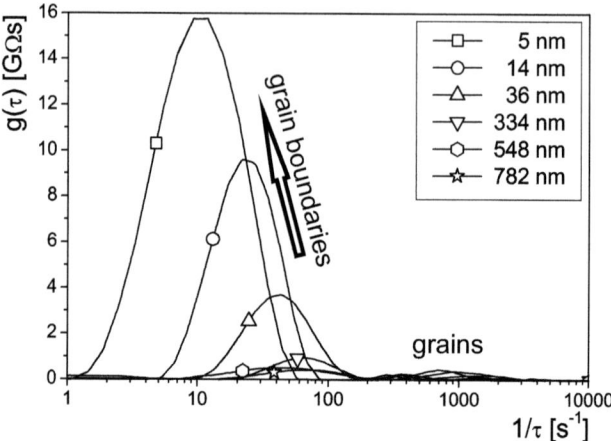

Figure 2.9: Distribution of relaxation times for the studied YDZ thin films determined from impedance spectra (measured at $T = 250\,°C$). The mean grain sizes of the specimens are given in the figure (cf. §3.3.1). [16, 167]

To elucidate the reason for the decrease of ionic conductivity upon decreasing mean grain size, impedance spectra of the YDZ thin films were analysed by calculating the distribution functions of relaxation times for each specimen. These distribution functions thus allow to distinctly separate the processes related to grain-boundary transport as well as the transport through the grains. Fig. 2.9 [16, 167] shows the distribution functions of relaxation times for the different specimens at $T = 250\,°C$. Two maxima are apparent contributing to the overall impedance. Whereas the pronounced maximum at lower frequencies has been attributed to the grain boundaries, the weak and nearly constant high-frequency process indicates the bulk transport through the grains.

Since the area enveloped by the curve is a measure for the resistance of the particular process, the spectra portend a nearly constant, grain-size independent contribution of the bulk resistance to the total resistance. The grain-boundary resistance, however, increases with decreasing mean grain size (arrow in Fig. 2.9) and causes an overall increase of the total resistance.

The shift of the grain-boundary maximum towards lower frequencies with decreasing mean grain size arises from the increasing overall grain-boundary resistivity. This is explained by the fact that the resonance frequency of this process is proportional to the grain-boundary conductivity, i.e., the reciprocal value of the resistivity.

2.2 Experimental Techniques

The goal of this work was the structural, microstructural, and chemical characterization of the specimens by means of scanning and transmission electron microscopy (SEM, TEM) techniques, while the electrical properties were examined at the *Institute of Materials for Electrical and Electronic Engineering IWE, University of Karlsruhe* as collaborating group. For further information, the reader interested in details about the characterization of the electrical properties, is referred to Ref. [15, 167] as well as to the PhD theses of C. Peters [16] and A. Müller [166].

Since numerous TEM techniques have been applied, the following section should only give a brief overview with emphasis on the most important aspects of the applied methods. Readers who are more interested in the materials aspects may pass over to chapter 3 (pp. 41).

2.2.1 Equipment

The microstructural characterization was carried out using a 200 keV Philips CM200 FEG/ST transmission electron microscope equipped with a field-emission gun. For bright-field and dark-field imaging, a 2k x 2k CCD camera from Tietz Video and Image Processing Systems GmbH (Gauting, Germany) was used.

Selected-area electron diffraction (SAED) patterns were recorded on imaging plates from Ditabis Digital Biomedical Imaging Systems AG (Pforzheim, Germany), providing a linear contrast transfer. In combination with a 16-bit scanner (Ditabis AG, Pforzheim, Germany) the resulting diffraction patterns exhibit a high dynamic range. A selected-area aperture with a diameter corresponding to about 200 nm was used for single-crystallite diffraction. To obtain Debye diffraction patterns of the nanocrystalline YDZ thin films, an aperture with a size corresponding to about $1\,\mu$m was used. Site-specific energy-dispersive X-ray spectrometry (EDXS) was carried out by means of a NORAN Vantage system with a Ge X-ray detector, using a probe diameter of about 2 nm.

Scanning TEM (STEM) was performed utilizing a 200 keV Zeiss LEO 922 microscope, equipped with an OMEGA energy filter. The high-angle annular dark-field (HAADF) detector in combination with the scanning unit allows to generate Z-contrast images and thus the visualization of the porosity within in the nanocrystalline YDZ thin films.

Energy-filtered TEM was in addition tried out to visualize the variations of the local Y and O contents that were found in 8YDZ, as will be shown later. But this technique failed due to several independent reasons, e.g. the small size of the chemical variations, the strong overlap of ionization edges at low energy losses, and the high energy losses of the $L_{2,3}$ ionization edges of the cations.

An aberration-corrected FEI Titan 80-300 with an image C_s-corrector was used to perform STEM in combination with analytical experiments (EDXS, EELS) on the nanoscale in order to correlate microstructural features and the local Zr, Y, and O concentrations in the thick-film electrolyte foils. The microscope was operated at 300 kV. The microscope was equipped with an EDAX X-ray detector operated at a dispersion of 20 eV/channel as well as a post-column energy filter (Gatan imaging filter, GIF) from Gatan GmbH (Germany).

STEM imaging using a HAADF detector was carried out using short camera lengths to visualize the regions, in which the analytical experiments were performed, at minimized dynamic conditions.

To correlate the position of coarsened tetragonal regions that were found in heat-treated 8YDZ with local chemistry, STEM imaging was performed using a long camera length of 301 mm. Since the inner collection angle of the HAADF detector is small at this long camera length, the {112} reflections are also detected. As a consequence, a high image intensity is obtained in regions with strongly excited {112} reflections (coarsened tetragonal regions). Such micrographs will be denoted as *dynamic* annular dark-field (ADF) images in the following.

In order to comprehensively study the local chemistry of the 8YDZ thick-film specimens on the nanoscale, both complementary spectroscopic techniques, namely energy-dispersive X-ray spectroscopy (EDXS) and electron-energy loss spectroscopy (EELS) were used simultaneously.

For simulating, e.g. diffraction patterns and images, the software package JEMS of P. Stadelmann [168] was used. Digital Micrograph from Gatan GmbH (Germany) was used for processing of the analytical data. The free peak-fitting software package Fityk [169] was used to apply empirical background subtraction to radial intensity distributions. Such intensity distributions were for example obtained from electron-diffraction patterns. This has become necessary because no reliable model was available in order to simulate the background contribution due to carbon films and/or contamination. Depending on the specific purpose, several software packages were used for image processing, i.e., ImageJ [170] and Adobe Photoshop CS4.

2.2.2 Quantitative EDXS

As will be shown in § 3.2.4, the Kα X-ray emission lines of the cations are applicable to determine the local cation contents from EDXS. After estimation and subtraction of a constant bremsstrahlung background (energy windows used for the background estimation: 11.8–14.3 keV) below the element-specific signals, the net intensities of the Kα lines of the cations (energy windows for Y: 14.62–15.22 keV and Zr: 15.42–16.02 keV) were used to determine the local Y and Zr content. These X-ray emission lines are well separable, while the L lines of Zr and Y strongly overlap (cf. Fig. 3.13, p. 62).

The errors introduced by approximating a linear background contribution in contrast to the correct modelling of the bremsstrahlung background is negligible in comparison to the statistical errors, because the experimentally measured background intensity itself is negligible in the investigated TEM sample regions.

A well-known procedure for the quantification of EDX data is the determination of k-factors, which relate the measured intensities and the chemical composition. According to Lorimer [171], the ratio of the atomic Zr and Y concentrations on the cationic sublattice, i.e. C_{Zr} and C_Y, can be determined by Eq. 2.3, in which the net intensities of the Kα lines of the cations are denoted as $I_{Zr\,K\alpha}$ and $I_{Y\,K\alpha}$. The factor $1/k_{ZrY} = ZAF$ as a product of the factors Z, A, and F correlates the measured intensities and the chemical composition. Z describes the direct production of X-rays by the impact of the primary electrons, which depends on the atomic number, as well as the energy-depending detection efficiency of the X-ray detector. F stands for secondary fluorescence due to the absorption (factor A) of primary X-rays in the TEM sample at finite sample thickness.

$$\frac{I_{Zr\,K\alpha}}{I_{Y\,K\alpha}} = 1/k_{ZrY} \cdot \frac{C_{Zr}}{C_Y} = ZAF \cdot \frac{C_{Zr}}{C_Y} \tag{2.3}$$

The X-ray mass attenuation coefficients [172] and thus the overall absorption of K X-rays of Y and Zr in YDZ itself is negligible. Hence, secondary fluorescence taken into account by the factor F in Eq. 2.3 can also be neglected. The small thicknesses (≤ 30 nm) of the investigated TEM sample regions, leading to a minimum of secondary effects, in addition justifies the following approximation, which simplifies the quantification. Therefore, Eq. 2.3 reduces to Eq. 2.4.

$$\frac{I_{Zr\,K\alpha}}{I_{Y\,K\alpha}} \approx Z \cdot \frac{C_{Zr}}{C_Y} = 1/k_{ZrY} \cdot \frac{C_{Zr}}{C_Y} \tag{2.4}$$

$$= \left(\frac{(Q\omega a)_{Zr}}{(Q\omega a)_Y} \cdot \frac{\epsilon_{Zr}}{\epsilon_Y}\right) \cdot \frac{C_{Zr}}{C_Y} \tag{2.5}$$

Only the factor Z is relevant for quantification in our case, which describes the production (Q_i: ionization cross-section, ω_i: fluorescence yield, $a_i = \frac{I_{i\,K\alpha}}{I_{i\,K\alpha}+I_{i\,K\beta}}$: yield for K$\alpha$ X-ray emission on the K shell, i = Zr, Y) and the detection (ϵ_i: detector efficiency, i = Zr, Y) of Kα lines.

The total K-shell ionization cross-section of Zr atoms for 300 keV impact is slightly smaller than that for Y atoms (Q_{Zr}=0.825 × 10^{-22} cm^2, Q_Y=0.883 × 10^{-22} cm^2, Ref.

[173]), while the Kα fluorescence yield for Zr, i.e., $\omega_{Zr} = 0.730$, is slightly larger than the value for yttrium ($\omega_Y = 0.710$) (Ref. [174, 175]). Further values are a_{Zr}=0.8394 and a_Y=0.8417 [176].

The efficiencies of the Si(Li)-detector ϵ_{Zr} and ϵ_Y can be assumed to be identical for the Kα lines of both elements. Calculating $1/Z$ finally amounts to $1/Z \approx 0.97$.

To verify the theoretically predicted value for $k_{ZrY} = 1/Z$, Eq. 2.3 was used to determine experimental values for the k-factor for the 8YDZ with well-known mean Y content from the obtained data sets. As will be shown in § 3.2.4, the experimental value, which has been obtained from reliable spectra of the as-sintered 8YDZ specimen with the quoted Y content of 8.52 mol%, arises to 1.02, which is slightly higher than the theoretically predicted value. Hence, the experimentally determined k-factor was used for the quantification of the local cationic composition.

As the sum of the Y and Zr concentrations on the cation sites amounts to 100 %, i.e., $C_{Zr} + C_Y = 1$, arithmetic transformation of Eq. 2.5 leads to the following equations (Eq. 2.6 and 2.7) that were used to determine the local Y and Zr content (at%) on the cationic sublattice from the spectral data.

$$C_Y = \frac{I_{Y\,K\alpha}}{I_{Y\,K\alpha} + 1.02 \cdot I_{Zr\,K\alpha}} \quad (2.6)$$

$$C_{Zr} = \frac{I_{Zr\,K\alpha}}{I_{Zr\,K\alpha} + 1/1.02 \cdot I_{Y\,K\alpha}} \quad (2.7)$$

Possible variations of the TEM sample thickness along a line scan do not affect the quantification of the cation concentrations by EDXS because each spectrum is quantified individually. Since all intensities, recorded within a single spectrum, are depending on the same illuminated volume, thickness correction is automatically achieved by dividing $I_{Y\,K\alpha}$ by the weighted sum of the Kα net intensities of both cations.

2.2.3 Quantitative EELS

Thickness Estimation

The absolute thickness of the studied TEM sample regions was estimated from EELS spectra. Since the probability of inelastic scattering of an incident electron increases with increasing sample thickness, it may undergo several independent inelastic scattering events. Due to this fact, multiple scattering obeys *Poisson* statistics. As a result, the fraction of inelastically scattered electrons exponentially increases with

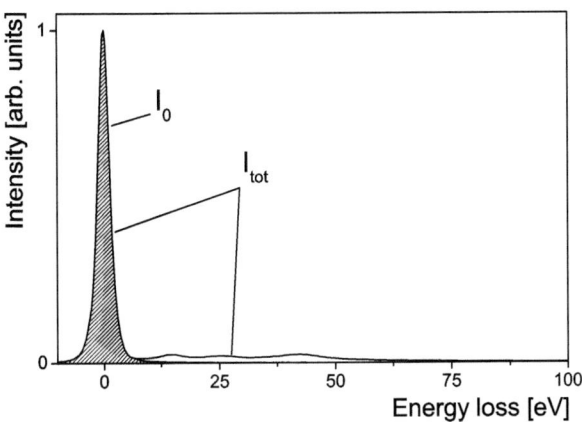

Figure 2.10: Experimental low-loss spectrum of 8YDZ-AS obtained at a relative thickness $\frac{t}{\lambda}=0.26$ corresponding to 31 nm. The symmetrical intensity contribution I_0 of the zero-loss peak is indicated.

increasing thickness.

$$I_0 = I_{\text{tot}} \cdot \exp\left(-\frac{t}{\lambda}\right) \qquad (2.8)$$

$$\frac{t}{\lambda} = \ln\left(\frac{I_{\text{tot}}}{I_0}\right) \qquad (2.9)$$

This is described by Eq. 2.8 where I_{tot} is the total number of electrons, i.e., elastically and inelastically scattered electrons, that are collected within a low-loss spectrum up to an energy loss of a few 100 eV. I_0 stands for the number of elastically scattered electrons in the zero-loss peak of the spectrum (cf. Fig. 2.10). Hence, the second equation (Eq. 2.9) was used to determine the local relative thickness. This value was used to judge the local sample thickness with respect to multiple scattering.

In this study, only sample regions with relative thicknesses in the range of $0.05 \leq \frac{t}{\lambda} \leq 0.3$ were investigated. This fact simplifies the evaluation of the EEL spectra since multiple scattering that may have a strong influence on quantitative EELS is negligible.

The inelastic contribution to the EEL spectra is dominated by plasmon excitations at energy losses below about 50 eV, within the exception of the case that element-specific ionization edges within the first 100 eV are present. This applies to the $N_{2,3}$ ionization edges of Y and Zr in this study. Since the spectral intensity rapidly decreases with increasing energy loss, integration within the first few 100 eV is sufficient for the determination of I_{tot}. I_0 has been determined by assuming a symmetrical zero-loss peak. While I_{tot} and I_0 can be derived from a low-loss spectrum (cf. Fig. 2.10) of the sample region of interest, the mean free path λ for inelastic scattering, i.e., the mean free path for mainly the excitation of plasmon-like resonances in the TEM sample, can be estimated by a parameterization proposed by Malis et al. [177]. Eq. 2.11 describes λ [nm] as a function of the effective atomic number of the studied material, i.e., Z_{eff}, the energy of the primary electrons E_0 [keV], and the collection semi-angle β [mrad] (cf. Table 3.2, p. 64).

$$\lambda = \frac{106 \cdot F E_0}{E_m \ln\left(\frac{2\beta E_0}{E_m}\right)} \tag{2.10}$$

with (2.11)

$$F = \frac{1 + E_0/1022}{(1 + E_0/511)^2} \tag{2.12}$$

$$E_m = 7.6 \text{ eV} \cdot Z_{\text{eff}}^{0.36} \tag{2.13}$$

$$Z_{\text{eff}} = \frac{\sum_i f_i Z_i^{1.3}}{\sum_i f_i Z_i^{0.3}} \tag{2.14}$$

The effective atomic number Z_{eff} is calculated (Eq. 2.14) by dividing the weighted sum of $Z_i^{1.3}$ by the weighted sum of $Z_i^{0.3}$ with f_i as atomic fraction of the elements $i = $ Zr, Y, O.

Using the known values for 8.5 mol% Y_2O_3-doped ZrO_2, the mean free path λ arises to 122 nm. This value is in good agreement with experimental values presented by Iakoubovskii et al. [178] who published values for λ, i.e., 115 nm for pure ZrO_2 and 122 nm for pure Y_2O_3.

The derived value on the basis of Eq. 2.9 has to be considered as an estimate for the local sample thickness in order to compare the thickness with microstructural features found in the specimens.

Background Modelling

Since the element-specific ionization edges are superposed to the falling background of ionization edges and the plasmon excitations at lower energy losses, each spectrum needs to be corrected for this background in order to extract quantitative element-specific signals. The background results from multiple inelastic scattering processes (plasmon excitation and ionization), which cannot be modelled in a deterministic way.

However, in energy-loss ranges sufficiently beyond preceding element-specific signals, the background intensity I_B can be estimated by fitting a mathematical expression, e.g. a power law ($I_B = A \cdot \Delta E^{-r}$) [179, 180]. Here, ΔE denotes the energy loss of primary electrons during propagation through the TEM sample. This function was used to determine the background below the Y-$M_{4,5}$ and Zr-$M_{4,5}$ as well as the Y-$L_{2,3}$ and Zr-$L_{2,3}$ ionization edges. To correct for the background below the O-K ionization edge, a 1st order log-polynomial law was chosen due to the better reproduction of the pre-edge background in comparison to the power-law fit.

Artefacts Induced by Background Correction

In the following, artificial variations in element-specific signal distributions along line-scan measurements are described, which were found to arise from modelling the background for each spectrum of the line-scan data set. The commonly applied procedure is to model (and subtract) the background for each spectrum of a data set by the same mathematical expression.

Using the example of the O-K ionization edge in 8YDZ, the development of such artefacts along a representative line scan of as-sintered 8YDZ is presented in Fig. 2.11. The displayed net intensity (Fig. 2.11a) was determined from the background-corrected spectra in an energy-loss range Δ of 50 eV starting with the edge onset at 532 eV energy loss. The factor N used for thickness correction (Fig. 2.11a, dashed line) was obtained from the post-edge net intensity in the range of 700–750 eV. While the correction factor only shows slight variations, implausible variations (jumps) of the net signal (exemplarily marked by the arrows) can be recognized. These variations are also present in the thickness-corrected O-K net intensity shown in Fig. 2.11b. To determine the origin of these variations, sequently recorded spectra (No. 21–25, thin lines in Fig. 2.12) around the untrustworthy evaluated spectrum No. 23 (thick line in Fig. 2.12, marked by arrows in Fig. 2.11) are plotted in Fig. 2.12. The raw spectra look similar except for small variations due to statistics, since no drastic change of the local sample thickness is observed (cf. factor N, Fig. 2.11a), which could explain the change of the net signal of the O-K ionization edge of spectrum No. 23.

The background contributions modelled by a 1st-order log-polynomial law are plotted for each spectrum. Despite the M_1 ionization edges of the cations (Y: 394 eV, Zr: 430 eV), which could if at all be faintly observed, the background was fitted using 100 eV wide energy-loss windows (365–465 eV). This window was chosen due to the broad but weak intensity maximum (500–530 eV) directly situated at lower energy losses prior to the O-K ionization edge (cf. Fig. 2.12a and b). The origin of this pre-edge intensity could not be clarified up to now. Nevertheless, this intensity was found in all recorded spectra of as-sintered and heat-treated 8YDZ.

As shown in the graph, the background estimation for the specific spectrum No. 23 fails (underestimation), although there is no necessity for a change of the slope of the background below the ionization edge from a physical point of view. One should know that the estimation of the background using a power law also failed for this spectrum. This failure explains the overestimation of the net signal (cf. lower curves in Fig. 2.12b) and thus of the integral O-K net signal for this measuring point of the

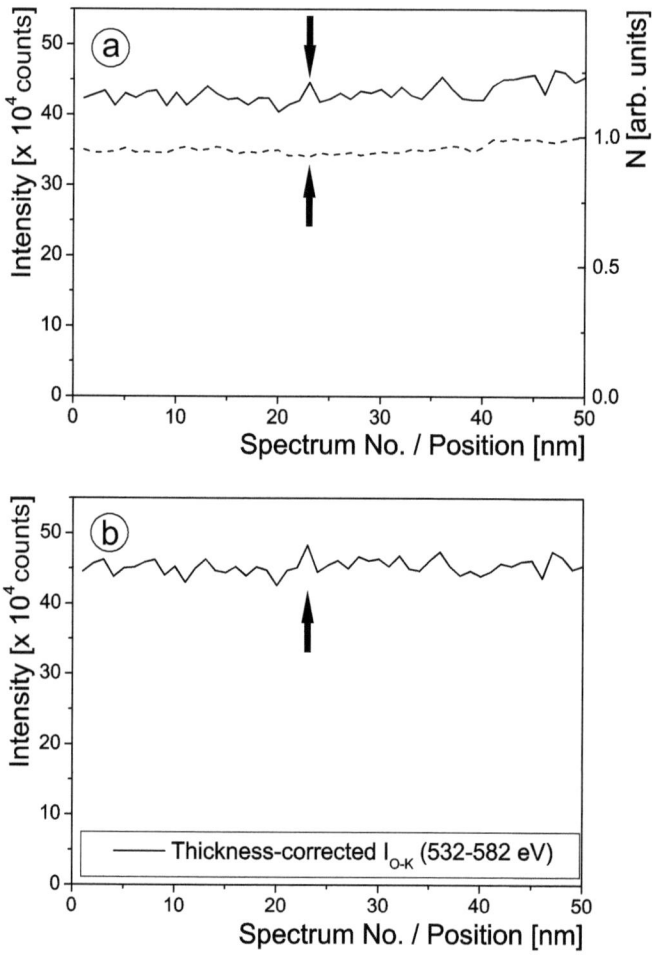

Figure 2.11: a) Integral net intensity (532–582 eV) of the O-K ionization edge (straight line) and corresponding factor N (dashed line) used for thickness correction along a line scan in 8YDZ-AS. N was determined from the integral intensity in the energy-loss range of 700–750 eV. b) Thickness-corrected O-K net intensity showing unrealistic variations of the O content. The black arrows exemplarily mark the conspicuous spectrum No. 23.

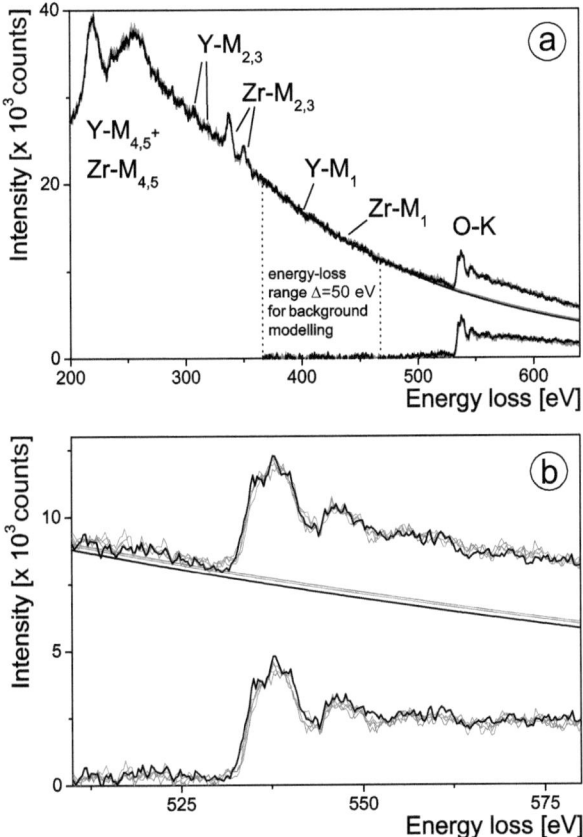

Figure 2.12: a) Superposed raw spectra No. 21–25 showing similar intensities, the modelled background (see text) is plotted below each spectrum. b) Energy-loss range 500–580 eV is shown magnified (thick line marks spectrum No. 23). In addition to the raw data, the background-corrected net signal of the O-K ionization edge is shown for each spectrum. After background subtraction, the net signal of spectrum No. 23 is overestimated.

line scan. The thick line marks the background-corrected spectrum No. 23 in average surpassing the other spectra.

Several reasons may be responsible for this problem:

- Statistics: Slight variations of the pre-edge background can lead to changes of the slope of the approximated background,

- Unstable background estimation using a power law or 1st order log-polynomial law: Slight variations of the pre-edge background may lead to strong variations below the ionization edge,

- Time-dependent sensitivity of the scintillator of the CCD camera after changing from low-loss to core-loss signals.

Such artefacts due to the background correction were found to arise during the evaluation of core-loss spectra in general. To minimize the influence of the background correction, an alternative evaluation technique was utilized in this work. Least-squares fitting of reference spectra to the raw data was found to be the much more reliable way to extract net signals from the analytical line scans.

Improvement by MLLS Fitting

In contrast to the commonly applied least-squares fitting of experimental data by an analytical expression by varying the parameters of this expression, the fitting of reference spectra to the experimental data was found to be the more reliable way to determine element-specific signals from EEL spectra. This procedure is similar to the fitting of an analytical expression except the fact that reference spectra with specific contours are used. Therefore, weighting factors $A, B, C, \ldots$ of the weighted sum of a set of reference spectra (background reference: R_{BG}, signal 1: R_{S1}, signal 2: R_{S2}, $\ldots$) are varied in such a way that the sum R^2 of the least squares of the differences $r(\Delta E_i)$ between measured values $I(\Delta E_i)$ and fit, i.e., the weighted sum of reference spectra in Eq. 2.16, become minimal (Eq. 2.17). Therefore, the energy loss ΔE_i of a specific channel i of a spectrum is used as the variable.

$$R^2 = \sum_i r(\Delta E_i)^2 \qquad (2.15)$$

$$= \sum_i \left(I(\Delta E_i) - [A \cdot R_{BG}(\Delta E_i) + B \cdot R_{S1}(\Delta E_i) + C \cdot R_{S2}(\Delta E_i) + \ldots] \right)^2 \qquad (2.16)$$

$$\frac{\partial R^2}{\partial A} \stackrel{!}{=} \frac{\partial R^2}{\partial B} \stackrel{!}{=} \frac{\partial R^2}{\partial C} \stackrel{!}{=} \ldots \stackrel{!}{=} 0 \qquad (2.17)$$

Two different procedures were utilized to obtain reference spectra for this evaluation procedure. First, reference spectra from well-defined reference materials can be measured, i.e., Y_2O_3 and ZrO_2, exhibiting similar contours of the element-specific ionization edges (near-edge structures) as the studied material. This was done in order to extract the net intensities of the Y-$L_{2,3}$ and Zr-$L_{2,3}$ ionization edges of YDZ for quantitative analyses (cf. §3.2.4).

A second method can be used, if only one non-overlapping element-specific ionization edge has to be studied and if the near-edge structures of this specific ionization edge do not vary significantly along a measured line scan. All spectra along the line scan are summarized to minimize statistical variations. From this sum spectrum, a reference background spectrum as well as a reference spectrum for the element-specific ionization edge were determined by fitting the background of the sum spectrum using an appropriate model (cf. §2.2.3, second paragraph). The *intrinsic* reference spectra were then used for MLLS fitting of each spectrum of the specific line scan. This procedure has been used to reliably extract the net intensity of the O-K ionization edge of YDZ as measure for the local oxygen content. As will be shown in the experimental part, this procedure yields significantly improved accuracy in determining the elemental distributions for the as-prepared material, which has been assumed to be chemically homogeneous, than presented in Fig. 2.11.

Since the procedure for MLLS fitting of reference spectra is implemented in Digital Micrograph, this software package was used for quantitative evaluation of the spectral data.

Quantitative EELS Analyses

Assuming single scattering of primary electrons passing through the TEM sample in sufficiently thin regions, a simple expression can be used to quantitatively evaluate element-specific net intensities. For this purpose the element-specific net intensities in given energy-loss windows Δ_A and Δ_B are determined by MLLS fitting, as outlined before.

Analogous to EDXS (Eq. 2.4 and 2.5), the connection between the local chemical composition and the measured intensities in EELS can be drawn (Eq. 2.18). In contrast to EDXS, only the probability for the excitation of a core-shell electron has to be considered in core-loss EELS. Hence, the measured intensity I_A for an element A only depends on the partial ionization cross-section $\sigma_A(E_0, \beta, \alpha, \Delta_A)$ and the number of atoms of this element in the probed sample volume (or the local atomic content C_A) [179]. Thus, for two elements A and B the following equation holds:

$$\frac{I_A(E_0, \beta, \alpha, \Delta_A)}{I_B(E_0, \beta, \alpha, \Delta_B)} = \frac{\sigma_A(E_0, \beta, \alpha, \Delta_A)}{\sigma_B(E_0, \beta, \alpha, \Delta_B)} \cdot \frac{C_A}{C_B} = 1/k_{AB} \cdot \frac{C_A}{C_B} \qquad (2.18)$$

The cross-sections σ_A and σ_B strongly depend on the primary electron energy E_0, the collection semi-angle β, the convergence semi-angle α, and the energy-loss window Δ. Δ describes the energy-loss window that is used for integrating the net signal of a specific ionization edge for quantitative analyses. It commonly starts at the onset of an ionization edge, if not otherwise specified. For a set of experimental parameters (E_0, β, α) in EELS, Eq. 2.19 was used to determine an experimental k_{AB}-factor [181] from reference materials or the as-sintered 8YDZ-AS specimen, i.e.

$$k_{AB} = \frac{I_B(E_0, \beta, \alpha, \Delta_B)}{I_A(E_0, \beta, \alpha, \Delta_A)} \cdot \frac{C_A}{C_B} \qquad (2.19)$$

For this study, quantitative EELS was mainly performed using the Y-$L_{2,3}$ and Zr-$L_{2,3}$ ionization edges. Assuming a completely filled cationic sublattice ($C_{Zr} + C_Y = 1$) leads to the same equations as in EDXS, i.e., Eq. 2.6 and 2.7, but with the EELS k-factor.

Due to the fact that the accuracy of theoretical partial cross-sections is strongly limited by the model that is used to describe ionization, i.e., the hydrogenic model [182] or the Hartree-Slater model [183–185], exclusively experimental k-factors were used in this study.

Chapter 3

Experimental results

3.1 Specimens

3.1.1 8YDZ Thick-Film Electrolytes

As-sintered and heat-treated 8YDZ electrolyte substrates (thickness $\sim 200\,\mu m$) with dopant contents of 8.52 mol% Y_2O_3 (8YDZ) and 10 mol% Y_2O_3 (10YDZ) were investigated to study the origin of the degradation of ionic conductivity in 8YDZ, as outlined in §2.1.1 (p. 8). The commercially available 8YDZ electrolyte substrates were fabricated by tape casting at Kerafol GmbH (Germany). For this purpose, pure powder from Toyo Soda Manufacturing Co., Ltd. (Japan) with quoted contents of impurities of Si < 30 ppm and Al < 10 ppm was used. In order to obtain gas-tight electrolyte substrates (relative density $\geq 98\,\%$), the green tapes were first heat-treated at 1300 °C for 5 h, followed by the final sintering at a temperature of 1550 °C for 2 h [166].
For microstructural comparison, one as-sintered 8YDZ substrate (8YDZ-ASH) was additionally homogenized at 1700 °C in the cubic phase field (cf. 2.1.2) for 4 h. The 10 mol% Y_2O_3-doped ZrO_2 electrolyte substrates were purchased from Nippon Shokubai Co., Ltd. (Japan). These substrates had been sintered at 1350 °C. All sintered YDZ substrates are polycrystalline with mean grain sizes on the microscale. The as-sintered specimens are named 8YDZ-AS and 10YDZ-AS, respectively.
Substrates with both dopant contents were annealed at 950 °C for periods of 2500 h (8YDZ) and 1000 h (10YDZ), respectively, to emulate the application as electrolytes in high-temperature SOFCs. The heat-treated specimens are denoted in the following as 8YDZ-H and 10YDZ-H. As outlined in §2.1.3, a decrease in ionic conductivity of nearly 40 % was found for 8YDZ, while the conductivity of 10YDZ was stable under the applied annealing conditions.
To study any influence of the electrical field, applied during DC characterization, on

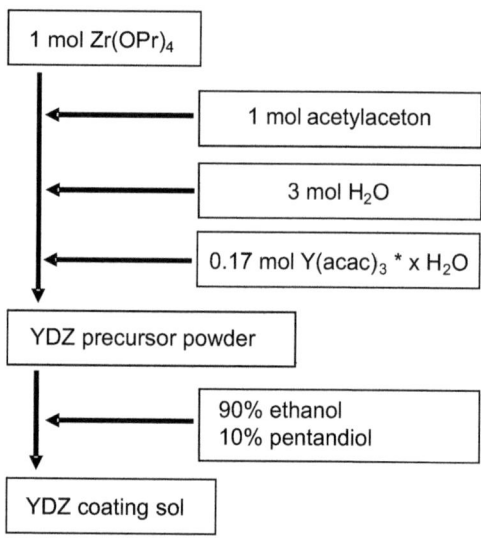

Figure 3.1: Chemical processing route of the coating sol used for the preparation of the YDZ thin films.

the microstructural coarsening of the 8YDZ-H electrolyte, a second series of long-term annealed 8YDZ thick-film electrolyte pieces was investigated. These specimens were heat-treated without electrical load at 1300 °C, 1250 °C, 1200 °C, and 1000 °C. Higher temperatures were used to reproduce the microstructural evolution on a shorter time scale. Also the position of the boundary between the cubic and the c+t two-phase field was of considerable interest for varying the annealing temperature. While the specimens heat-treated at ≥ 1200 °C were annealed for 150 h (1300 °C and 1250 °C) and 600 h (1200 °C), the specimens at 1000 °C were heat-treated for 1500, 3000, and 5800 h, respectively. All these electrolytes were heat-treated under air to reproduce the microstructural coarsening, which was found in 8YDZ.

The specimens heat-treated at 1000 °C are denoted as 8YDZ-L1500, 8YDZ-L3000, 8YDZ-L5800, depending on the annealing time, while the electrolytes heat-treated at the higher temperatures are denoted together as 8YDZ-LHT. All specimens are marked in the schematic phase diagram presented in Fig. 2.2.

3.1.2 Nanocrystalline YDZ Thin Films

For thin-film preparation, a synthesis route based on soluble precursor powders was utilized. Therefore, a Y_2O_3-doped ZrO_2 precursor powder with a ratio Zr/ligand/H_2O of 1/1/3 was prepared as follows (cf. Fig. 3.1): By slow addition of 1.0 mol chelate ligand to 237.7 g (1.0 mol) zirconium(IV)-propoxide a yellow sol was obtained, which was stirred for one hour. After cooling down to room temperature, 54.0 g (3.0 mol) deionised water and 58.8 g (173 mmol) yttrium acetate hydrate were added. The sol was stirred for 30 minutes at room temperature followed by 30 minutes at 80 °C. Without further delay, subsequently all volatile components were removed from the reaction mixture by rotational evaporation at reduced pressure (< 40 mbar) with a maximum bath temperature of 90 °C. The coating solution was prepared with an oxide yield respective to crystalline 8YDZ of 6 mass% by dissolution of the amorphous precursor powder in a solvent mixture of 90 mass% ethanol and 10 mass% 1,5-pentanediol. After stirring over night, they were filtered through a 0.45 μm membrane [186, 187].

Thin films were prepared by dip coating of isolating r-sapphire substrates (Crystec, Germany, 0.53 mm × 52.5 mm × 52.5 mm, one side polished, average roughness $R_a = 0.06$ nm), used without further cleaning. Thin-film preparation was carried out in clean room atmosphere under constant humidity conditions of 20 % relative humidity at 24 °C (4.4 g H_2O/m^3) using a climate control unit. The substrate was dipped in the dip-coating container and the solution was let to settle down for 15 seconds. The coatings were made by ten times multiple dip-coating combined with a withdrawal rate of 30 mm/min. A standard drying time of 2 min was used for each sol-gel layer. The calcination of each single-layer sol-gel film was performed by rapid thermal annealing (RTA) at 500 °C with a dwell time of 10 min. Subsequently, the specimens were exposed to a final tempering step at temperatures in the range of 650–1600 celsius for 24 h respectively in order to adjust mean grain size and degree of porosity. The samples were prepared using heating ramps of 5 K min^{-1} (≤ 1000 °C) or 20 K min^{-1} (> 1000 °C) [188, 189].

Depending on the temperature of the final tempering step, the different samples are named YDZ-650, YDZ-850, YDZ-1000, YDZ-1250, YDZ-1350 and YDZ-1600 in the following. All YDZ specimens studied in this work are summarized in Table 3.1.

3.1.3 Y_2O_3 and ZrO_2 References for Analytical TEM

Two reference materials, namely monoclinic zirconium dioxide (ZrO_2) (quoted impurity contents by weight: $HfO_2 < 2$ %, $SiO_2 < 50$ ppm, $Al_2O_3 < 50$ ppm) and yttrium oxide (Y_2O_3) (quoted contents of impurities by weight: Zr < 430 ppm, Si < 80 ppm) nanoparticles, purchased from Treibacher Industrie AG (Austria), were analyzed to obtain reference spectra for quantitative EELS.

Fig. 3.2 shows TEM bright-field images of both types of nanoparticles, which were separately deposited on Lacey films from Plano GmbH (Germany) by nebulizing dispersions of particles in ethanol by means of an ultrasonic vaporizer. The utilized

Specimen	Y$_2$O$_3$ content [mol%]	Preparation T [°C] (t [h])	Additional heat treatment T [°C] (t [h])
8YDZ-AS	8.5	1550 (2)	-
8YDZ-ASH	8.5	1550 (2)	1700 (4)
8YDZ-H	8.5	1550 (2)	950 (2500)
8YDZ-L1500	8.5	1550 (2)	1000 (1500)
8YDZ-L3000	8.5	1550 (2)	1000 (3000)
8YDZ-L5800	8.5	1550 (2)	1000 (5800)
8YDZ-LHT	8.5	1550 (2)	1200-1300 (150-600)*
10YDZ-AS	10	1350 (4)	-
10YDZ-H	10	1350 (4)	950 (1000)
YDZ-650	7.3	650 (24)	-
YDZ-850	7.3	850 (24)	-
YDZ-1000	7.3	1000 (24)	-
YDZ-1250	7.3	1250 (24)	-
YDZ-1350	7.3	1350 (24)	-
YDZ-1600	7.3	1600 (24)	-

Table 3.1: List of YDZ electrolyte specimens investigated in this study. The highest temperatures, the specimens were exposed to, are indicated (*cf. §3.1.1).

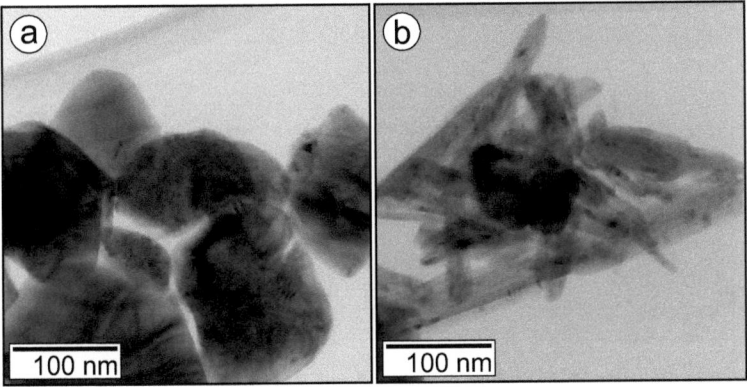

Figure 3.2: Bright-field TEM images of the a) ZrO_2 and b) Y_2O_3 reference nanoparticles on Lacey films.

technique using ethanol as solvent necessitates the cleaning of the TEM samples prior to the analytic experiments with a plasma cleaner from Binder Labortechnik (Germany) to remove residues of the solvent.

The shape of the purchased Y_2O_3 particles (Fig. 3.2b) is needle-like with a length of a few 100 nm and a width of less than 50 nm, while the ZrO_2 nanoparticles (Fig. 3.2a) are spherical with $d_{50} < 2\,\mu m$.

The expected crystal structures of both reference materials at room temperature were verified by means of selected-area electron diffraction. From the phase diagram (cf. §2.1.2), the monoclinic phase is expected for pure ZrO_2, while pure Y_2O_3 should be cubic. Fig. 3.3 depicts Debye electron-diffraction patterns including reflections of numerous nanoparticles resulting in concentric rings of reflections. Background-corrected radial intensity scans of these diffraction patterns are presented in Fig. 3.4 in comparison to simulated data. To correct the radial intensity scans with respect to the diffuse background, the background below the reflections was empirically estimated and fitted by splines, using the free software tool Fityk [169]. The simulations using JEMS [168] are based on the structural data given in §2.1.2 (Table 2.1, p. 15). The crystal structures of both types of nanoparticles correspond well with the expected ones (from simulation) within the limits of electron diffraction. The disagreement of the relative intensities of measured and simulated reflections may be caused by the preferred alignment, especially of the needle-like Y_2O_3 nanoparticles with respect to the Lacey film.

The use of films with holes yields particles which protrude the Lacey film. Recording spectra from such regions has minimized or even prevented background intensities caused by the carbon film.

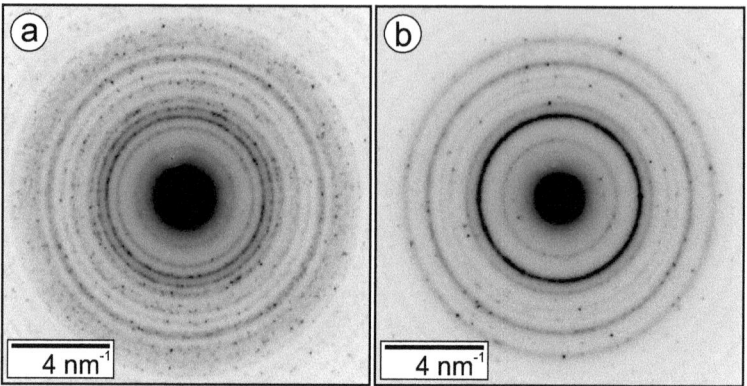

Figure 3.3: Debye electron-diffraction patterns of the a) ZrO_2 and b) Y_2O_3 reference nanoparticles.

The small size of the nanoparticles provides regions with small thicknesses ($t/\lambda \leq 0.25$, cf. Eq. 2.9, p. 33) with respect to the inelastic mean-free paths for plasmon excitations, resulting in excellent reference spectra for EELS — spectra with negligible to undetectable contributions of multiple scattering. The spectra for the detectable element-specific ionization edges, obtained from these references, are discussed in detail in § 3.2.4.

Despite the difference in crystal structure and atomic configuration compared to the investigated YDZ electrolyte specimens, the near-edge structures of the M and L ionization edges are very similar, as will be shown in § 3.2.4. Since the M and L ionization edges of the cations Y and Zr overlap, the reference data was used to separate the net intensities of the element-specific edges for Y and Zr by multi linear-least square fitting these reference spectra to the experimental data.

3.1.4 Reference Specimens for Raman Spectroscopy

Since the studied YDZ specimens exhibit cubic as well as tetragonal structure two single-phase specimens were used as reference materials for Raman spectroscopy. To evaluate the original spectra, reference spectra from a cubic as well as a tetragonal-type material are indispensable. Therefore, a 3.5 mol% YDZ (3YDZ) thick-film electrolyte and a 15.6 mol% YDZ (16YDZ) single crystal were used. The 3YDZ thick-film electrolyte was prepared by tape-casting. The crystal form was mainly t'-YDZ due to the rapid cooling (cf. § 2.1.2, p. 20). The 16YDZ single crystal was purchased from Zrimat Corp. (North Billerica, USA). The crystal structure is exclusively cubic [15].

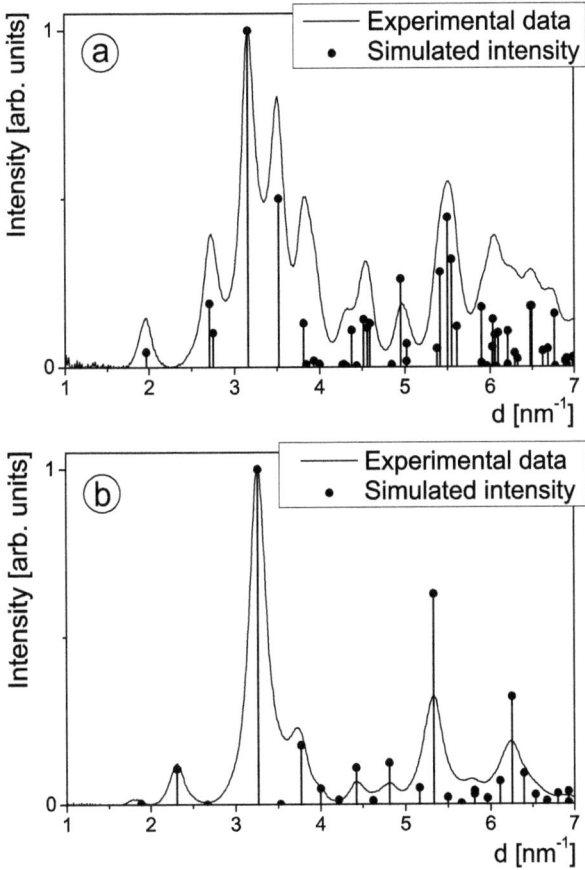

Figure 3.4: Background-subtracted radial intensity distributions of the diffraction patterns of a) ZrO_2 and b) Y_2O_3 nanoparticles (Fig. 3.3) in comparison to simulated intensity distributions based on the structural parameters given in Table 2.1 (p. 15).

3.1.5 Sample Preparation for TEM

Two different techniques were utilized to prepare samples for TEM investigation. Bulk as well as cross-section samples of dense microcrystalline 8YDZ and 10YDZ electrolytes were prepared by common procedures including disk-cutting, grinding, dimpling, polishing and Ar^+-ion milling. The thickness of the samples after polishing was about a few µm. This yields short ion-milling durations at incidence angles between 10° and 16° and an ion current of 1 mA (Gatan Duomill). For the final polishing, the acceleration voltage of 4 kV was reduced to 2.5 kV. The TEM samples prepared for structural and microstructural investigation, for which homogeneous illumination is used in the microscope, were coated with a thin layer of carbon to reduce charging effects and thus sample drift in the microscope. The thin region in the middle of the samples was masked in order to minimize artefacts due to the carbon coating. For analytic investigation using a fine electron probe, a few TEM samples were cleaned using a plasma cleaner from Binder Labortechnik (Germany) immediately before the experiments to minimize or possibly even prevent sample contamination.

The preparation of cross-section samples of the YDZ thin films by standard procedures, especially of those specimens annealed at temperatures above 1250 °C, occasionally led to cracking of the single-crystalline sapphire substrates. In these cases, TEM lamellae of the thin films with homogeneous thicknesses between 50–150 nm were prepared by focused ion beam (FIB) milling with a SEM/FIB dual beam system (EsB 1540 Crossbeam, ZEISS, Germany). Therefore, mainly the H-bar technique [190] was utilized. After preparation, a thin layer of carbon was deposited on the samples except the thin region of interest in order to reduce charging effects in the electron microscopes.

3.2 8YDZ Thick-Film Electrolyte

The optimized sintering procedure for the preparation of the purchased 8YDZ and 10YDZ substrates results in homogeneous and gas-tight polycrystalline electrolyte foils with mean grain sizes of a few µm. No secondary phases due to the segregation of impurities were detected, neither by direct imaging techniques like high-resolution TEM nor by analytical TEM (EDXS) [15]. Since changes of the bulk of the polycrystalline material had been expected from impedance spectroscopy, the focus of the following part is on microstructural and chemical changes within the grains and regions in the vicinity of grain-boundaries.

3.2.1 Crystal Structure

It is known from electron-diffraction analyses (cf. 2.1.2, p. 23) that even at 8.5 mol% YDZ is not fully stabilized in the cubic high-temperature phase. The diffraction pattern of a single grain of as-sintered 8YDZ, oriented in [111]-ZA, in Fig. 3.5 shows the superposition of tetragonal-type and cubic reflections. While the dominating reflections are due to both types of structures, the additional weak reflections can clearly be attributed to tetragonal symmetry. The identification of this type of structure, taking double diffraction as artefact in slightly too thick sample regions into account, is discussed in detail in a recent publication by Butz et al. [15]. For simplification, the cubic indexing scheme is used in the following.

While the indicated {112} reflections in Fig. 3.5 are kinematically allowed for any tetragonal-type phase (t-, t'-, t"-YDZ), the inner ring of 6 weak {110} reflections, not marked in Fig. 3.5, is due to multiple scattering. The pairwise disappearance of opposing {110} reflections indicates the presence of three variants of tetragonal phase with the c-axis aligned parallel to all three cubic main axes. This disappearance of forbidden reflections as indication for the tetragonal nature has been shown along various higher indexed directions in thin sample regions [191].

At the positions of the strong reflections in Fig. 3.5 tetragonal-type reflections with large structure factors superpose the cubic ones. Therefore, the kinematically allowed but weak {112} reflections had to be used for the imaging of the tetragonal phase.

Despite low structure factors (simulated using the structural data given on p. 15) and thus low intensities of these additional reflections, independent of the nature of the tetragonal phase (t-, t'-, t"-YDZ), the distribution of the tetragonal phase has been visualized by applying dark-field TEM imaging using the intensity of a tetragonal {112} reflections.

Therefore, single crystallites was commonly oriented with an ⟨111⟩ or ⟨110⟩ axis of the cubic phase parallel to the incident electron beam. After slightly tilting out zone

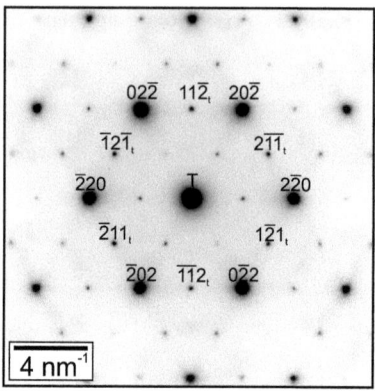

Figure 3.5: SAED pattern of 8YDZ-AS along the [111]-ZA showing weak additional reflections next to the cubic-type reflections, the mainly cubic {220} and tetragonal-type {112} reflections are indicated.

axis (ZA) to minimize the excitation of most of the strong reflections and thus to enhance the excitation of the {112} reflection chosen for the imaging process, this additional reflection was centred on the optical axis. For dark-field imaging, this centred reflection was selected, using the smallest objective aperture of the microscope.

3.2.2 Phase Distributions

The resolved distribution of the tetragonal phase in 8YDZ-AS is depicted in Fig. 3.6. The field of view includes a single grain at the hole of the TEM sample oriented close to the [111]-zone axis. This hole of the TEM sample appears black in the lower left corner of the image. In such {112} dark-field micrographs, the tetragonal phase appears with bright contrast, while the cubic matrix is dark.
Homogeneously distributed, nanoscaled tetragonal precipitates are observed in 8YDZ-AS, which are coherently embedded in the cubic matrix. No change of the density of these precipitates was detected in the vicinity of grain boundaries.
A similar microstructure as in as-sintered 8YDZ, homogeneously precipitated tetragonal nanoscaled regions, is found in 10YDZ, but with a smaller density of precipitates of the tetragonal phase [15]. Even 8YDZ-ASH exhibits similar tetragonal precipitates in spite the additional homogenization in the cubic phase field at 1700 °C.

Annealing 8YDZ at 950 °C for 2500 h (8YDZ-H) leads to a distinct change of the microstructure, as depicted in Fig. 3.7a, where a significant coarsening of the tetragonal regions is observed. This coarsening was reproduced by the series of the specimens 8YDZ-L1500, 8YDZ-L3000, and 8YDZ-L5800, where 8YDZ-L3000 showed a similar

Figure 3.6: Dark-field micrograph obtained using the intensity of a kinematically allowed {112} reflection to visualize the distribution of the nanoscaled tetragonal precipitates in 8YDZ-AS.

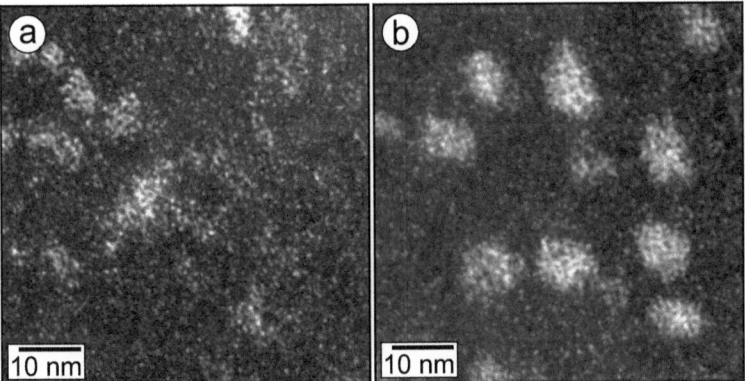

Figure 3.7: Dark-field micrographs of a) 8YDZ-H and b) 8YDZ-L5800. The microstructural coarsening is characterized by the local agglomeration of the tetragonal nanoscaled precipitates in spherical regions with sizes of about 10 nm.

microstructure in comparison to 8YDZ-H, despite the slightly higher temperature of the heat treatment (1000 °C instead of 950 °C). In contrast, the specimens, which were heat-treated at temperatures exceeding 1000 °C (denoted as 8YDZ-LHT in Fig. 2.2, cf. 42) did not show any microstructural changes with time. The microstructure of these specimens is the same as compared to the as-sintered 8YDZ material.

This microstructural evolution further develops during heat treatment at 1000 °C as shown in Fig. 3.7b. The spherically-shaped coarsened regions in 8YDZ-L5800 (Fig. 3.7b) appear in a more advanced stage, which means more distinct but with similar size compared to 8YDZ-H (Fig. 3.7a) and 8YDZ-L3000. It can be recognized in Fig. 3.7 that the coarsened regions in 8YDZ-H and 8YDZ-L5800 rather consist of agglomerated nanoscaled precipitates than of precipitates with a homogeneous constitution of tetragonal symmetry.

In contrast, no significant change of the microstructure was observed in 10YDZ-H heat-treated at 950 °C for 1000 h. The distribution of the tetragonal phase in 10YDZ-H appears similar to that in 10YDZ-AS [15].

To study the temporal evolution of the microstructure of the specimens during TEM investigation, the exposure time for dark-field imaging could be reduced from 1–2 s [15] to 50–250 ms by applying 4-fold hardware binning on the CCD chip. This necessitated the increase of the magnification to assure a sufficiently high lateral resolution. Since the installed slow-scan CCD camera needs about 250 ms for the read-out of the detector, the temporal resolution is limited to about 3 images per second. Series of consecutive images were taken to investigate time-dependent microstructural changes that had not been noticed yet.

In addition to the *simple* detection [15], the temporal evolution of the location of the bright contrast corresponding to the tetragonal precipitates has become recognible under common illumination conditions using the Philips CM200. In 8YDZ-AS and 10YDZ-AS, the location of the tetragonal nanoscale precipitates (cf. Fig. 3.6) was noted to fluctuate within the grains. This is visualized for 8YDZ-AS in Fig. 3.8 by the series of chronological dark-field micrographs obtained by intervals of about 300 ms. The life/dwell time of the tetragonal nanoscale regions in the investigated region is on the timescale of a few images, which means on the time scale of only few seconds. With the available temporal detection limitation, any movement of the regions could not have been resolved. Thus, it cannot be distinguished between the appearance/disappearance and the movement of the precipitates within the grains.

Since the microstructure of 10YDZ is similar to that of as-sintered 8YDZ but with a lower density of tetragonal precipitates the same phenomenon, the temporal instability of the tetragonal precipitates, can be observed in 10YDZ-AS and 10YDZ-H.

The temporal evolution of 8YDZ-L5800 during the investigation in the TEM is depicted in the series of chronological dark-field micrographs in Fig. 3.9. As observed

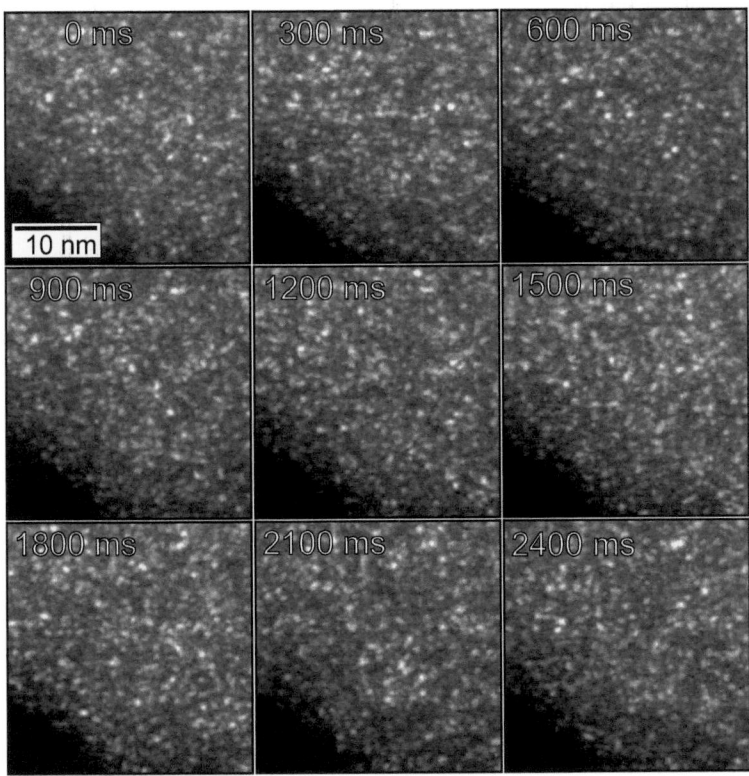

Figure 3.8: Time series of dark-field micrographs of 8YDZ-AS (same sample region).

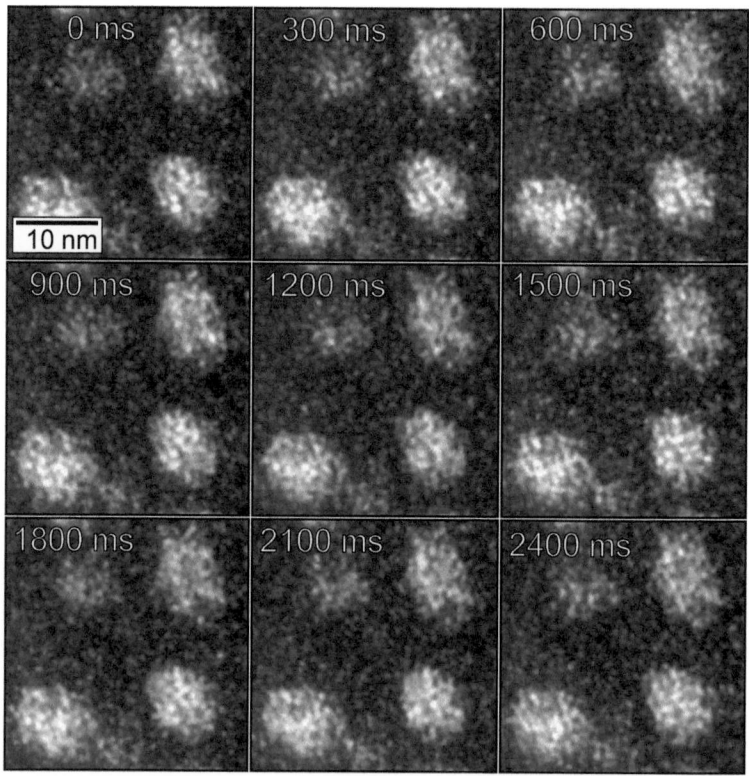

Figure 3.9: Time series of dark-field micrographs of 8YDZ-L5800 (same sample region).

for as-sintered 8YDZ, the location of the bright contrast corresponding to tetragonal precipitates, which are observed in between the coarsened regions, fluctuates as well. In contrast, the agglomerated precipitates in the coarsened regions of 8YDZ-L5800 remain almost statically but separated. This difference in stability of the tetragonal precipitates in the material indicates a difference in stability of the formed phase depending on the location in the specimen.

The influence of the incident electron beam on the observed phenomenon was tested by reducing the current density, the gun parameters and spreading of the illumination to a minimum for the observation of the precipitates themselves. The rate of the fluctuations could in addition be reduced by the use of a liquid-nitrogen sample holder in the TEM. Nevertheless, the fluctuation of the tetragonal regions in 8YDZ-AS were observed as well.

At present, it is not possible to distinguish if the fluctuations are induced by electron-beam irradiation, i.e., local heating and/or charging, or if it is an intrinsic property of the materials. However, the fluctuations indicate the instability of the observed tetragonal regions — irrespective of the cause.

3.2.3 Volume Fraction of Tetragonal YDZ

For the discussion of the observed microstructural evolution of 8YDZ it is meaningful to get an idea of the volume fraction of the non-cubic regions in the investigated YDZ specimens. Therefore, two complementary methods were tried out. The indirect method by analysing the intensities measured in electron diffraction turned out to be inappropriate to reliably determine any volume ratio of tetragonal to cubic YDZ. This is mainly caused by the fact that the structure factors especially of the nanoscaled tetragonal precipitates are not known satisfyingly and may in addition vary within the precipitates. Also, dynamic scattering depending on local sample thickness and size of the tetragonal regions may contribute to the uncertain determination of the volume fraction using the weak additional reflections.

As direct method, the evaluation of the distribution of the tetragonal precipitates from dark-field images was utilized. This procedure failed for the homogeneous distributions of nanoscaled precipitates observed in as-sintered 8YDZ, 10YDZ-AS, and 10YDZ-H, because the regions are only separable in extremely thin regions due to the contrast overlap in regions thicker than approximately 10 nm.

Nevertheless, the procedure turned out to be suitable to determine limits for the volume fraction of the coarsened regions in 8YDZ-H and 8YDZ-L5800 since these regions appear as separable regions up to local TEM thicknesses of about 50 nm. For the evaluation, a wedge-shaped lamella of 8YDZ-L5800 with a given angle of 15° was prepared by FIB Ga-ion milling. This sample consisted of a few grains. One of the

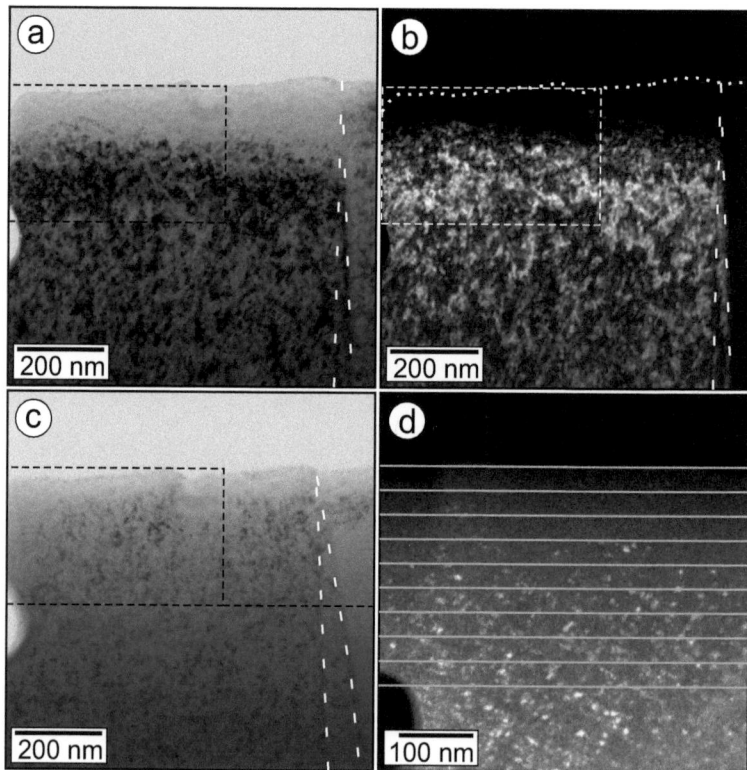

Figure 3.10: a) Bright-field micrograph and b) corresponding dark-field micrograph of wedge-shaped FIB lamella of 8YDZ-L5800 oriented in $\vec{g} = (\bar{1}11)$ two-beam condition close to the [110] ZA. Orientation is close to the surface normal of the sample.
c) Bright-field micrograph of 30° tilted sample. The projection of grain boundary in the field of view appears broadened in comparison to a). d) Dark-field micrograph using tetragonal $(\bar{1}12)$ intensity to visualize the coarsened mainly tetragonal regions in 8YDZ-L5800. The dashed rectangle in the images marks the sample region, which was evaluated to determine the volume fraction of the coarsened regions.

grains was intrinsically oriented with the surface normal close to [110]. To validate the wedge angle, bright-field and corresponding dark-field micrographs of this grain in $\vec{g} = (\bar{1}11)$ and $\vec{g} = (002)$ 2-beam conditions were analysed with respect to the thickness contours due to dynamic scattering. The bright-field and corresponding dark-field image of the grain with $\vec{g} = (\bar{1}11)$ is presented in Fig. 3.10a and Fig. 3.10b. Despite the strong strain contrast in both images, one can estimate the location of thickness contours by averaging the image intensities parallel to the edge of the TEM sample as plotted in Fig. 3.11b. The extinction lengths $\xi_{(000)} \approx 45$ nm and $\xi_{(\bar{1}11)} \approx 50$ nm were obtained from simulations using JEMS [168]. Therefore, the amplitudes of the transmitted and the $(\bar{1}11)$ beam depending on sample thickness were calculated by Bloch-wave and multislice calculations for the defined orientation of the TEM sample. The dependency of these values on the local thickness is plotted in Fig. 3.11. While discrepancies can be recognized between the absolute values of the amplitudes calculated by both methods, the positions of the minima and maxima of the corresponding curves, marked as vertical lines in Fig. 3.11, are nearly the same for the Bloch-wave and the multislice simulations.

The local sample thickness of one extinction length is reached in a distance d of about 300 nm from the edge of the lamella, as indicated in the averaged intensity distributions of the bright-field and the dark-field image in the evaluated sample region (cf. Fig. 3.11). From these images, the estimated wedge-angle, which was calculated by $\alpha = 2\arctan\left(\frac{\xi_{\vec{g}}/2}{d}\right)$, resulted in $9°\pm 2°$.

The field of view presented in Fig. 3.10 includes a slightly tilted grain boundary, which is marked by the dashed white lines for better visualization. This grain boundary was used for a second test of the local thickness, since the projection of this tilted grain boundary broadens when the sample is tilted with the tilt axis in the grain-boundary plane. Fig. 3.10c shows the same region of the sample after tilting 30°. As can be recognized, the width of the grain boundary projection increased. This increase of the width of the projection was additionally used to estimate the local sample thickness. Both methods yielded a thickness of well over 50 nm in a distance of nearly 300 nm from the sample edge.

The discrepancy between nominal (15°) and measured ($9°\pm 2°$) wedge angle was expected due to the small thickness with respect to the size of the interaction volume of the Ga ions during FIB milling. Thus, sputtering has to be expected on both surfaces of the wedge, resulting in an increased milling rate in thinner sample regions.

Fig. 3.10d shows the dark-field image using the intensity of the tetragonal $(\bar{1}12)$ reflection visualizing the coarsened tetragonal regions. The projected size of the evaluated region (square marked by dashed lines in Fig. 3.10a-c) is 300 x 500 nm^2. In order to determine the volume fraction, this marked region was sectioned into 9 fields, as indicated in dark-field micrograph of Fig. 3.10d, each 32 nm wide.

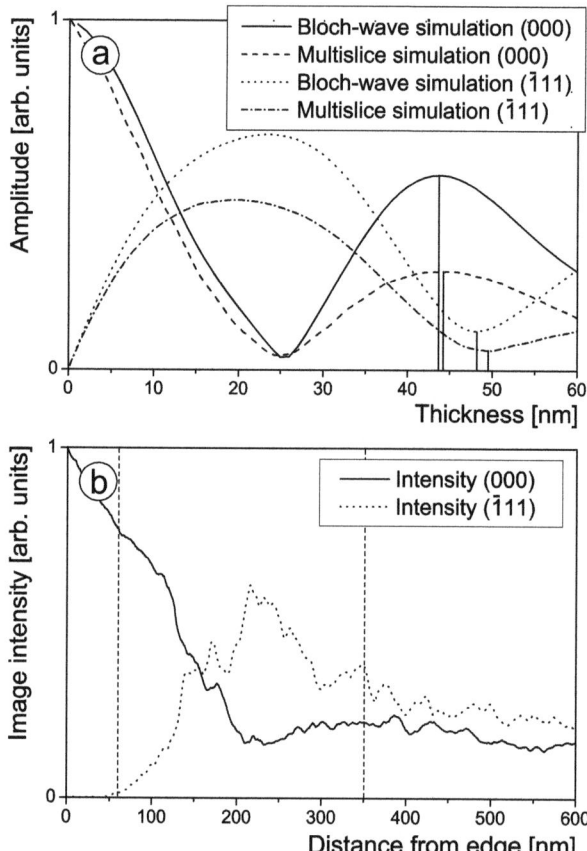

Figure 3.11: a) Calculated amplitude of (000) and ($\bar{1}11$) reflections (Bloch-wave and multislice simulations) depending on thickness in the same orientation as the images in Fig. 3.10a and b were taken. b) Averaged (parallel to the sample edge) intensity distributions of the bright-field and dark-field images presented in Fig. 3.10a and b.

A few assumptions, which were made for determining a lower and an upper limit for the volume fraction, are mentioned in the following. Most of all, one has to take into account the truncation of the spherical regions at the surfaces of the TEM sample during ion etching. With decreasing thickness, an increasing fraction of the identified regions is not completely situated in the TEM sample.

- Only coarsened regions, which were clearly identified, were counted in each field

- These numbers include truncated (up to 50 %) regions situated at both surfaces of the wedge-shaped sample

- Coarsened regions, which were truncated more than 50 %, could ocularly not be detected in the dark-field micrograph of the FIB lamella that was taken at intermediate magnification

A lower limit for the volume fraction was determined by multiplying the counted number of regions with the factor $P_I = (t - 2d)/t$ (t local average thickness in the specific field, d diameter of coarsened regions $\sim$4.5 nm). The factor P_I stands for the probability that the detected regions are completely embedded in the TEM sample and not truncated. This correction ensures that detected but truncated coarsened regions are not taken into account for the determination of the lower limit.

To determine an upper limit, both the truncated regions, which were detected as well as all regions, which were not recognized, have to be added to this value. Therefore, the number of the detected but truncated regions is calculated by multiplying the number of counted regions by the factor $P_T = 2d/t$.

The number of truncated regions, which were not detected arises to the same number. The truncated regions that are mainly in the specimen and those that are mainly truncated complement each other. Thus, this calculated number has to be used to determine the volume of truncated coarsened regions.

The determined value for coarsened regions which are situated within the FIB lamella resulted in 244, while the upper limit for all regions including the truncated ones arises to 292. The volume of one coarsened region with a mean radius of 4.5 nm arises to 382 nm^3.

Since only one of the three variants of the tetragonal phase is visualized in Fig. 3.10d, this is when the incident electron beam is perpendicular to the c-axis of the tetragonal phase, the factor 3 has to be taken into account when calculating the volume fraction of the coarsened tetragonal regions. Taking the uncertainties of the final wedge angle into account, the limits for the calculated volume fraction arise to 12 vol% and 7.5 vol%. Hence, the estimated volume fraction of tetragonal coarsened regions in 8YDZ-L5800 is about 10±3 vol%.

3.2.4 Chemistry on the Nanoscale

Due to the complexity of the phase diagram, the discussion about the microstructural coarsening of 8YDZ during heat treatment necessarily requires the examination of the chemical evolution on the nanoscale. If the microstructural decomposition of 8YDZ into the cubic and tetragonal YDZ phase has indeed progressed in 8YDZ-H and 8YDZ-L5800 in comparison to 8YDZ-AS and 8YDZ-ASH, variations have to be expected of the local Y and O concentrations on the scale of the agglomerated precipitates with a reduction of the Y concentration in the coarsened regions.

While the microstructure regarding the distribution of the tetragonal-type phase was determined for the as-sintered and heat-treated 8YDZ and 10YDZ specimens, as discussed in the previous part, the chemical composition with respect to the local oxygen as well as Zr and Y contents was extensively studied by means of EDXS and EELS only for 8YDZ before and after heat treatment.

The cation concentrations were measured quantitatively by EDXS and EELS, while EELS is more appropriate to monitor variations of the local oxygen content by the analysis of the intensity of the O-K ionization edge. To validate the EDXS results, local variations of the EELS intensities of the Y-$L_{2,3}$ and Zr-$L_{2,3}$ ionization edges were, in addition, quantitatively evaluated.

Since the coarsened regions embedded in the grains of 8YDZ-H are of sizes below about 10 nm (Fig. 3.7), the analytical experiments had to be conducted in thin TEM sample regions with thicknesses in the range of a few nm to 30 nm. Such thin regions turned out to be most suitable to monitor local chemical variations in the heat-treated 8YDZ specimen (8YDZ-H) on the scale of the size of the coarsened regions. A second important advantage is that multiple scattering, which occurs at larger thicknesses, could be neglected, evaluating the EELS data in this study.

Systematic analytical line-scan experiments with the simultaneous acquisition of EEL and EDX spectra were performed parallel to the edge of wedge-shaped TEM sample regions of as-sintered 8YDZ and 8YDZ-H at different distances to the sample edge. This procedure allows the assumption that the TEM sample thickness is nearly constant along each line. Fig. 3.12 depicts HAADF STEM images of representative sample regions of 8YDZ-AS and 8YDZ-H, in which systematic measurements were carried out. Both micrographs were recorded immediately after the experiments. Thus, the locally reduced sample thickness along the lines due to electron-beam induced damaging is visible in both images. The cloudy contrast in Fig. 3.12b results from local charging of the TEM sample in the imaging mode.

The length of each line scan was set to 50 nm, including 50 measuring points (1 point per nm). The dwell time per point was 20 s (core-loss EELS) or 5 s (low-loss EELS),

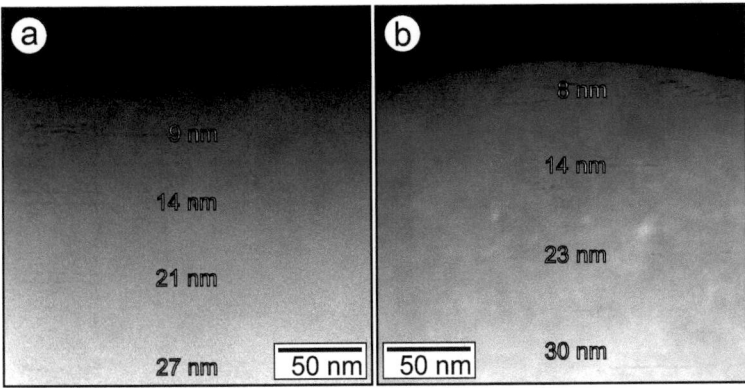

Figure 3.12: HAADF STEM images of representative wedge-shaped TEM sample regions of a) 8YDZ-AS and b) 8YDZ-H, in which analytical line-scan experiments were performed systematically. The given local thicknesses were estimated from EELS by the evaluation of low-loss spectra, taking the mean free path for plasmon excitation into account (cf. § 2.2.3).

depending on the energy-loss range and the illumination parameters which were optimized in order to maximize the collected intensity in EELS (see Table 3.2). Drift correction during a line-scan acquisition was performed each 10 spectra. Hence, the overall exposure time per line scan was about 20 minutes (core-loss EELS).

In the following, some experimental aspects of quantitative EDXS and EELS, which emerged during data evaluation, are summarized, using the results of the example of as-sintered 8YDZ. Readers who are more interested in the analytical results regarding the changes of the chemical composition of 8YDZ before and after the heat treatment may skip to § 3.2.4 on page 80.

Qualitative EDXS

In EDXS, all element-specific X-ray emission lines of oxygen as well as the cations are observed in the recorded spectra, as shown representatively in Fig. 3.13.

Due to the relatively short dwell time, the small TEM sample thickness in the studied regions and the small solid angle covered by the EDX detector, the overall X-ray intensities are relatively low (Fig. 3.13), resulting in large statistical variations compared to simultaneously recorded EELS data.

Despite the low detector efficiency at low X-ray energies, the O-K emission line is clearly visible at 0.525 eV. Due to the insufficient energy resolution of about 85 eV at 2 keV, the series of L X-ray emission lines of yttrium and zirconium cannot be

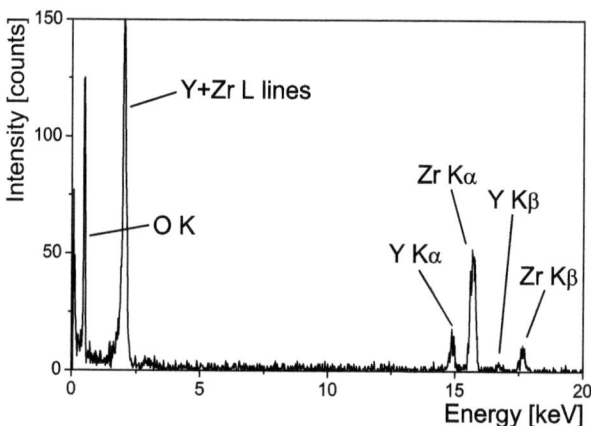

Figure 3.13: Representative EDX spectrum of one of the analytical line-scans (local sample thickness 14 nm, acquisition time as used for line-scan experiments 20 s) of as-sintered 8YDZ, showing all element-specific emission lines.

resolved as single lines. In contrast, the Kα and Kβ emission lines of Zr and Y at higher energies can clearly be separated as shown in Fig. 3.13. The background intensity due to bremsstrahlung is negligible at these high energies. Nevertheless, the background below the K emission lines of the cations was linearly interpolated and subtracted in order to determine the element-specific net intensities.

Qualitative EELS

An overview about the characteristic energy losses of the primary electrons in 8YDZ including the low-loss region and the element-specific ionization edges is given in Fig. 3.14.

The electron configurations of Zr and Y atoms are [Kr] $4d^2.5s^2$ and [Kr] $4d^1.5s^2$, respectively. Assuming oxidation states of 4+ for Zr ions and 3+ for Y ions in 8YDZ, similar as in the reference materials, both Zr^{4+} and Y^{3+} have the same electron configuration as [Kr]. This means, the K, L, and M shells are filled completely, while only the 4s and 4p states of the N shell are filled with electrons. Therefore, the complete series of K, L, and M ionization edges can be observed for both cations. In contrast, only transitions from the $4s_{1/2}$, $4p_{1/2}$, and $4p_{3/2}$ states of the N shell can be observed (N_1, N_2, and N_3 ionization edges).

Since the spectral intensity in EELS dramatically decreases with increasing energy loss, for better visibility the intensities in the different energy-loss ranges in Fig. 3.14 are depicted with a gain of 500 and 10^4, respectively.

The initial intention was to simultaneously record the M ionization edges of the

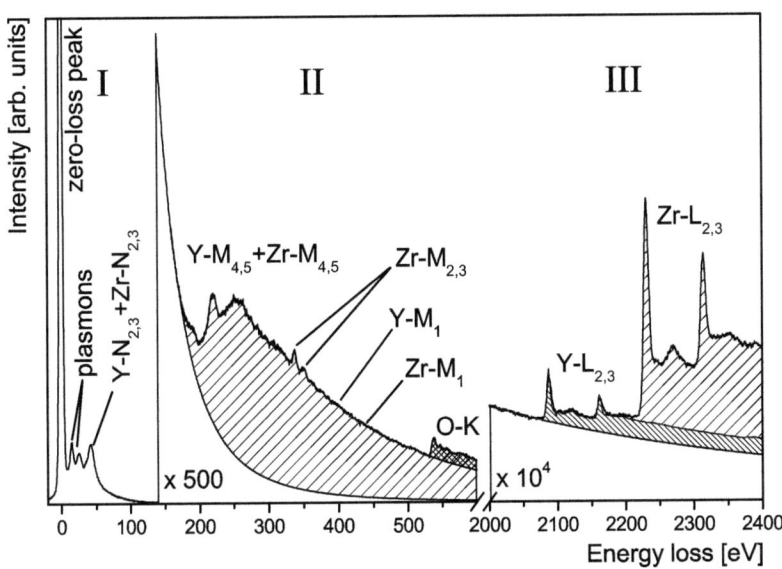

Figure 3.14: Overview about the detectable electron energy-loss features of 8YDZ including the zero-loss peak followed by the plasmon peaks and the element-specific ionization edges up to 2400 eV. The spectrum covers the three energy-loss ranges I, II, and III as indicated.

	Energy loss [eV]	Condenser aperture [μm]	Camera length [mm]	Convergence semi-angle α [mrad]	Collection semi-angle β [mrad]	Dwell time [s]
Low-loss	0-500	50	301	11.1	1	5
Y-M, Zr-M series, O-K	120-730	50	160	11.1	3.7	20
Y-L, Zr-L series	1860-2470	150	77	33.3	19.2	20

Table 3.2: Experimental parameters for quantitative EELS using the FEI Titan microscope.

cations as well as the O-K ionization edge (energy-loss range II in Fig. 3.14) for quantitative analyses. Therefore, the dispersion of the energy filter was set to 0.3 eV per channel for all EELS measurements. This leads to a theoretically covered energy window of about 615 eV, using the 2k x 2k CCD camera of the GIF (2048 channels). EEL spectra in the three different energy-loss ranges depicted in Fig. 3.14 were recorded to observe all relevant ionization edges of Y, Zr, and O, respectively.

Low-loss spectra with the elastically scattered intensity (zero-loss peak) were obtained to estimate the local sample thickness according to Eq. 2.9.

Spectra in the energy-loss range of 120–730 eV (cf. Fig. 3.14, energy-loss range II) were recorded, which include the M ionization edges of both cations as delayed maxima as well as the O-K ionization edge at 532 eV.

To record the $L_{2,3}$ ionization edges of Y (2080 eV) and Zr (2222 eV), spectra were acquired in the energy-loss range of 1860–2470 eV (cf. Fig. 3.14, energy-loss range III).

The experimental settings for the used FEI Titan microscope, including condenser aperture, camera length, convergence and collection semi-angles as well as dwell time for the different energy-loss ranges, are summarized in Table 3.2.

In the following, representative spectra of as-sintered 8YDZ are presented for the recorded energy-loss ranges. Reference spectra obtained from the Y_2O_3 and ZrO_2 nanoparticles (cf. §3.1.3) at similar sample thicknesses are plotted for comparison. These reference spectra were recorded using the same experimental conditions as used for the line-scan experiments (Table 3.2).

Since no significant changes of the near-edge fine structures of the element-specific ionization edges were observed along the analytical line scans using an energy resolution of 1–2 eV (depending on the experimental conditions), the comparison between

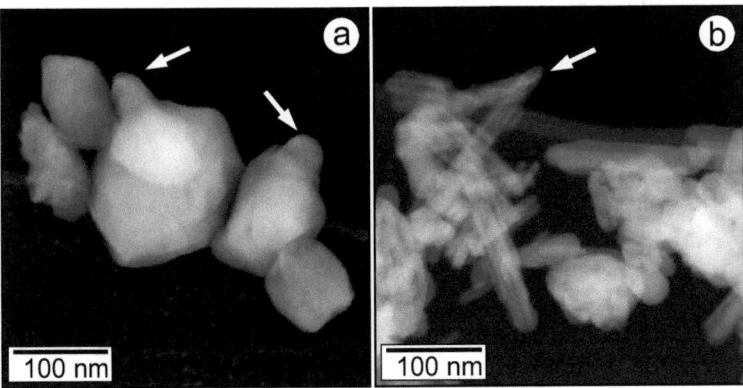

Figure 3.15: HAADF STEM images of a) ZrO_2 and b) Y_2O_3 nanoparticles, both used as references for the quantitative evaluation of EEL spectra of 8YDZ.

the spectra in different spectral energy-loss ranges of 8YDZ with spectra of the reference materials will only be outlined in a qualitative way.

To prevent spectral intensity caused by the carbon film, onto which the nanoparticles were deposited, the reference spectra were taken from nanoparticles protruding the carbon Lacey film. HAADF STEM images of such regions, from which the reference spectra were taken, are presented in Fig. 3.15 (see arrows). Since the cross-section of the thin Lacey film for scattering into large angles is very low, the film itself is only faintly observable in the images without gamma correcting the raw data.

In the low-loss region of 8YDZ, two pronounced peaks at 14.4 eV and 25.2 eV, respectively, due to plasmon excitations are measurable (Fig. 3.16). The positions of the maxima correspond to the energy losses due to plasmon excitation observed in the monoclinic ZrO_2 nanoparticles. The Y_2O_3 reference spectra on the other hand only show one pronounced plasmon peak with an energy loss corresponding to the first peak of ZrO_2 at 14.4 eV. The most obvious difference between 8YDZ and the reference materials is the more pronounced second plasmon peak at about 25.2 eV.

The third maximum at about 42 eV in the low-loss spectra of 8YDZ is mainly due to the $N_{2,3}$ ionization edge of the zirconium ions. A slight bump on the rising side of this edge, which is attributed to the overlapping intensity of the $N_{2,3}$ ionization edge of the yttrium ions, can be recognized comparing the curve of 8YDZ obtained at 25 nm with the spectrum obtained from the ZrO_2 reference at a similar thickness. It has not been possible to separate element-specific signals for quantitative EELS in this energy-loss range up to now.

The shapes of the M ionization edges of pure ZrO_2 and Y_2O_3 appear similarly as delayed maximum with the M_5 ionization edge (peak), followed by a pronounced

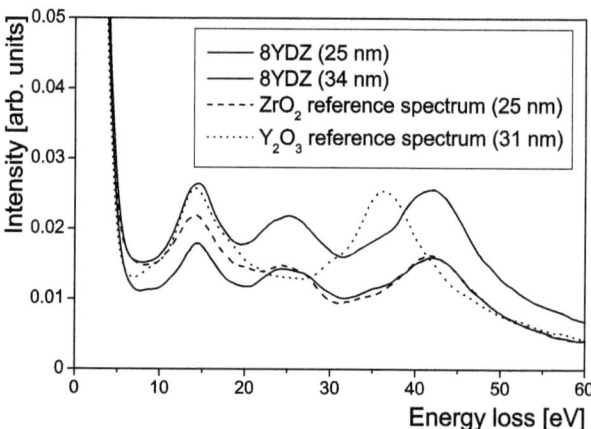

Figure 3.16: Low-loss EEL spectra of 8YDZ at 25 nm and 34 nm, low-loss spectra of both reference materials.

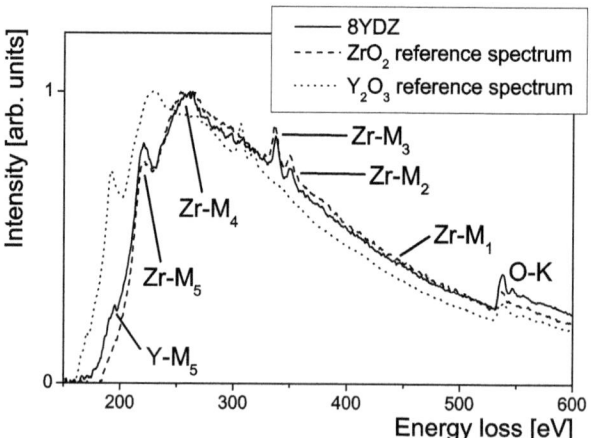

Figure 3.17: Background-subtracted EEL spectra of ZrO_2, Y_2O_3, and 8YDZ in energy-loss range II including $M_{4,5}$ ionization edges of Y and Zr as well as the O-K ionization edge at 532 eV.

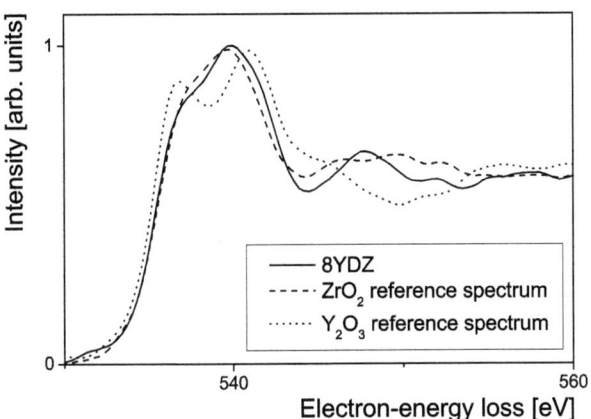

Figure 3.18: Background-subtracted O-K ionization edge of ZrO_2, Y_2O_3, and 8YDZ at 532 eV energy loss.

broadened peak due to the M_4 ionization edge. The weak but sharp $M_{2,3}$ peaks are situated on the falling background of the $M_{4,5}$ ionization edge, while the M_1 ionization edges, which should be observable as abrupt onsets at 394 eV (Y) and 430 eV (Zr), are only vaguely detectable. The different features of the M ionization edges are marked exemplarily for Zr in Fig. 3.17.

Since the $M_{4,5}$ ionization edges of Zr and Y strongly overlap and since the given Y content is only 17 at%, in this energetic range the EEL spectra of 8YDZ are mainly determined by the Zr ions. Nevertheless, a weak maximum can be faintly observed at the rising edge of the Zr-$M_{4,5}$ ionization edge, indicating the presence of Y in the material. This maximum corresponds to the Y-M_5 ionization edge.

In the spectra the O-K ionization edge is present at 532 eV as weaker feature in comparison to the M ionization edges of the cations (cf. Fig. 3.17). For better visibility this region is magnified in Fig. 3.18. In Fig. 3.17 it can be recognized that the jump ratio, i.e., the ratio of the O-K signal and the background, in the spectra is different despite the same parameters used for the acquisition. This has mainly been attributed to the different stoichiometries of ZrO_2, Y_2O_3, and 8YDZ.

Fig. 3.18 shows the near-edge fine structures of the O-K edge of 8YDZ in contrast to those of the two reference materials in detail. Any energetic shift of the edge onset could not be resolved within the detection limits.

While this ionization edge clearly shows a double-peak structure for the Y_2O_3 nanoparticles with two maxima at 536.8 eV and 541 eV, respectively, only one asymmetric maximum at about 539.8 eV is found for the ZrO_2 nanoparticles at the used energy resolution. A similar near-edge fine structure of the O-K ionization edge as for ZrO_2

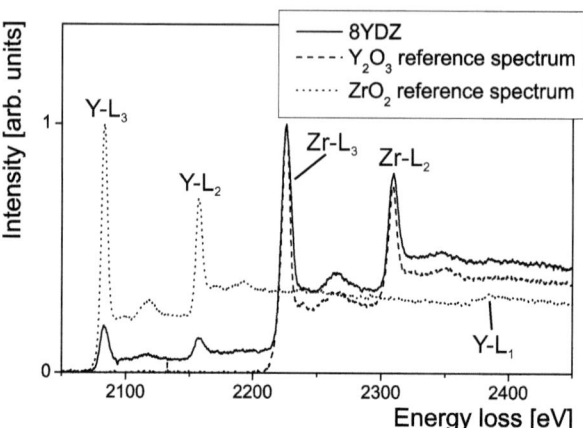

Figure 3.19: Background-subtracted Zr-L$_{2,3}$ and Y-L$_{2,3}$ ionization edges of ZrO$_2$, Y$_2$O$_3$, and 8YDZ.

is observed in 8YDZ within the first 10 eV from the edge onset (532–543 eV). A second distinct maximum is observed at 547.6 eV for 8YDZ, while a few weaker maxima are present in the near-edge structures of the reference spectrum of pure ZrO$_2$. The similarity of the spectra obtained from 8YDZ and the ZrO$_2$ nanoparticles is explained by the dominance of Zr as the main cation in 8YDZ. The differences at higher energy-losses arise from differing coordination in the monoclinic ZrO$_2$ in comparison to the mainly cubic structure of 8YDZ.

The Zr-L$_{2,3}$ and Y-L$_{2,3}$ ionization edges of the reference materials as well as of 8YDZ are shown in Fig. 3.19. As the M$_{4,5}$ ionization edges, the L$_{2,3}$ edges of Zr and Y exhibit similar features but at different energy losses. The most pronounced features of these ionization edges are the sharp white lines at the onsets of each ionization edge (L$_3$, L$_2$). These white lines are marked for the individual edges of Y and Zr (Fig. 3.19).

Homogeneity of As-Sintered 8YDZ

As outlined in §2.2.2, the Kα emission lines of the cations (cf. Fig. 3.13) were used to quantitatively investigate the local Y and Zr contents along the measured line scans. The k-factor k_{ZrY} of 1.023±0.007 was determined by summing up at least 500 line-scan EDX spectra of as-sintered 8YDZ, obtained at sample thicknesses in the range of 10–30 nm. This procedure yielded a negligible statistical error of the factor. Fig. 3.20 shows the quantified Y content of two representative line-scan experiments within a series of parallel line scans of 8YDZ-AS, obtained at thicknesses of 9 nm and

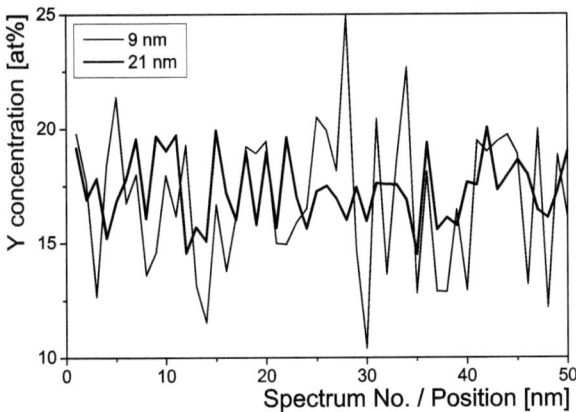

Figure 3.20: Y content (EDXS) along two representative line scans of 8YDZ-AS obtained at thicknesses of 9 nm and 21 nm.

21 nm, respectively, simultaneously recorded with the EELS data in the energy-loss range II.

As the thickness and thus the recorded intensities in the spectra increase, the variations of the Y content along the line scans decrease. The mean value of the Y content within each line scan is shown in Fig. 3.20 (measurements belong to line-scan series 1). The values of line-scan series 2 and 3 were evaluated from further systematic measurements recorded simultaneously with EEL spectra using energy losses in the ranges II and III.

The evaluated mean Y contents of all three series of line scans (Fig. 3.21a) coincide well with the given dopant content of 17 mol % on the cationic sublattice corresponding to 8.5 mol % Y_2O_3 at local thicknesses larger than 10 nm.

Only in thinnest regions below about 10 nm, being inappropriate for analyses due to predominant electron-beam induced damage as discussed in §2.2.2, the Y content appears to be underestimated.

In Fig. 3.21b, the experimental deviations of the local Y contents (1σ) from the mean values within each line-scan experiment of series 1 (cf. Fig. 3.21b) are presented. These deviations are marked by thin cap lines. To judge the strong variations of the Y content regarding chemical inhomogeneities in the as-sintered material, the statistically expected standard deviations arising from error propagation due to the low intensities in EDXS are plotted as thick cap lines for comparison. These values are slightly but systematically smaller than the experimentally obtained ones. Up to now, the reason for this systematic disagreement has not been found.

To clarify any chemical inhomogeneity on the cationic sublattice in the as-sintered

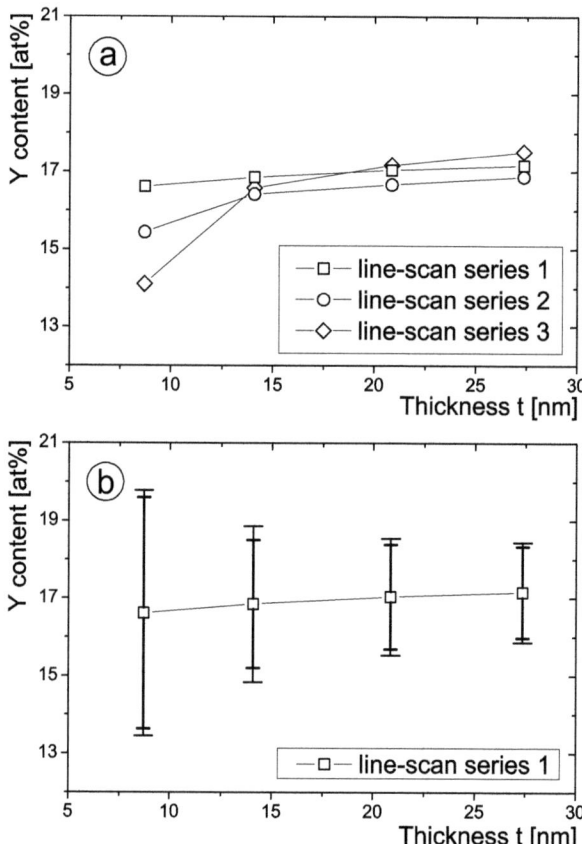

Figure 3.21: a) Mean values of the Y content in representative line-scan series of 8YDZ-AS. b) Experimental deviation from the mean value (thin cap lines) in comparison to the theoretically expected standard deviation (thick cap lines) within line-scan series 1. In both figures, the lines between the measuring points are guides to the eyes to associate related line-scan experiments of a studied region.

8YDZ, EEL spectra including both Y and Zr ionization edges were analysed regarding the local cationic composition.

Quantitative EELS

As summarized in Fig. 3.14, different element-specific intensity contributions were recorded simultaneously, depending on the evaluated energy-loss range.

Two different procedures were used to evaluate the experimental data regarding chemical composition and homogeneity along the experimental line scans. On the one hand, k-factors, which make the connection between net intensities of different elements were used to directly quantify the local composition for each single spectrum along the line scans. In this case, any correction for the local thickness is not necessary as discussed for quantitative EDXS (cf. §2.2.2).

The second option is the evaluation of the homogeneity of the net signal attributed to a specific element along the line scans. This procedure obviously necessitates the correction with respect to the local sample thickness since the thickness slightly varies along the line scans, as will be shown in the following.

It turned out during evaluation of the spectral data in the energy-loss ranges I and II (cf. Fig. 3.14) that both the N as well as the M ionization edges of the cations have been inappropriate up to now to determine the local cation contents. This is caused by the relatively low Y content and therefore the lower signal contribution associated with this element, which is superposed by much strong intensity contributions due to the Zr ions. In the following, the quantitative evaluation of the Zr-$L_{2,3}$ and Y-$L_{2,3}$ ionization edges in energy-loss range III using both above-mentioned procedures will be described in detail.

In energy-loss range II, only the net signal of oxygen was analysed with respect to variations of the anion content in as-sintered 8YDZ and 8YDZ-H along the measured line scans, using the second outlined procedure.

L Ionization Edges In energy-loss range III (cf. Fig. 3.14), the subsequently following L ionization edges of the cations were simultaneously recorded with the EDX spectra of line-scan series 2 (cf. Fig. 3.21). The four line scans parallel to the sample edge were taken at thicknesses in the range of about 10–30 nm.

After determination of the net signals of Y and Zr by MLLS fitting the reference spectra to the experimental data, the local composition on the cationic sublattice was evaluated applying the outlined procedures. First, the thickness-correction using a correction factor N extracted from an integral background intensity or the MLLS-fit parameter for the background for each spectrum will be described. Secondly, the results of the direct quantification are presented.

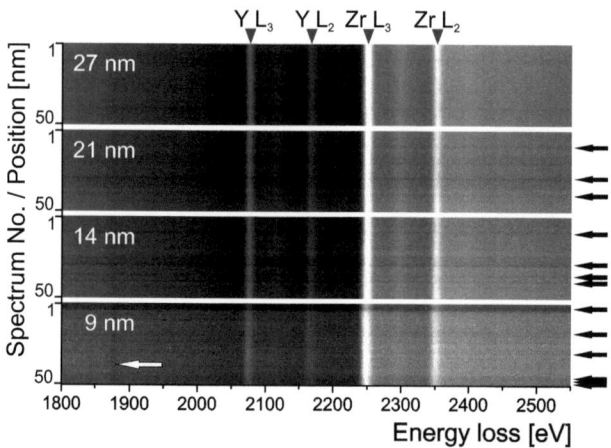

Figure 3.22: Visualization of a representative series of line scans of 8YDZ-AS, obtained in energy-loss range III at local thicknesses of 9 nm, 14 nm, 21 nm, and 27 nm. Each image represents the grey-scale coded spectra of a line scan from the first (top) to the last (bottom) recorded spectrum as visualized in *Digital Micrograph*. The bright lines correspond to the $L_{2,3}$ white lines of Y and Zr, respectively.

The four separate images in Fig. 3.22 show the set of four recorded line scans of 8YDZ-AS in the energetic region of 1800–2550 eV, as visualized in *Digital Micrograph*. Each image stands for a data set of 50 spectra, chronologically recorded (each 1 nm) from the top to the bottom along a single line scan. The estimated local sample thickness was about 9 nm, 14 nm, 21 nm, and 27 nm, respectively.

The L_3 and L_2 white lines of the ionization edges are apparent as white lines in the gray-scale images. It can be noticed that noise within the line-scan data sets decreases with increasing sample thickness, since the signal intensity increases linearly with increasing thickness.

The slight decrease in the local thickness for single spectra within a line scan appears as reduced line intensity in the image. Black arrows on the right-hand side of the graph mark spectra with ocularly reduced intensity within the line scans.

The slightly reduced intensity at about 1880 eV (marked by the white arrow), which can be seen in all line scans, is explained by switching from zero-energy loss (zero-loss peak) to the core-loss region, causing the locally varying sensitivity of the scintillator of the GIF CCD camera.

As described above, the L ionization edges of Y and Zr partly overlap. MLLS fitting (§ 2.2.3) of each individual spectrum of the experimental sets of spectra within a line scan by the Zr-$L_{2,3}$ and Y-$L_{2,3}$ reference spectra (Fig. 3.19) was utilized to extract the element-specific signal contributions.

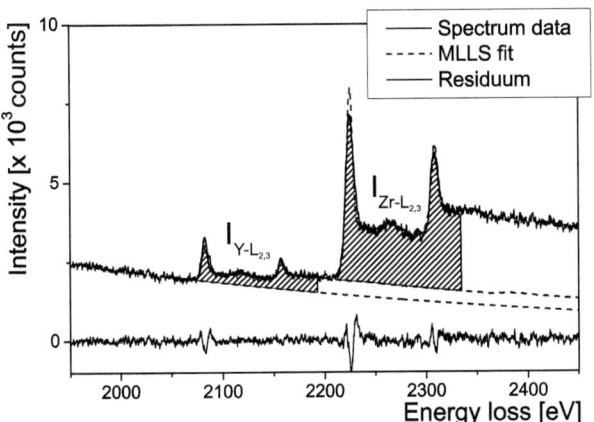

Figure 3.23: Representative MLLS fit of a single 8YDZ spectrum. The signal contributions, i.e., background, Y-$L_{2,3}$ ionization edge, and Zr-$L_{2,3}$ ionization edge are separated by dashed lines. The residuum between experimental data and fit is plotted below the spectrum. The evaluated integral intensities $I_{Zr-L_{2,3}}$ and $I_{Y-L_{2,3}}$ are marked by shaded areas.

The background reference spectrum, which was used for the evaluation of a line-scan data set, was obtained from the sum spectrum of all spectra of this specific line scan by fitting a power-law function in the energetic range of 1860–2060 eV. This assures optimum consideration of experimental settings and conditions on the background reference spectrum. Finally, the reference spectrum was normalized to the highest intensity at lowest energy loss in the spectrum.

In the following, the evaluation of the line scan obtained at 14 nm (Fig. 3.22) is described exemplarily in detail. For all other data sets, only the most important results will be presented.

Fig. 3.23 shows the representative MLLS fit of the reference spectra applied to the randomly chosen spectrum No. 17 of the line-scan data set obtained at 14 nm sample thickness (Fig. 3.22). In the energy region with strongly overlapping element-specific intensities (energy losses >2080 eV), each specific contribution of the MLLS fit related to Y or Zr is marked separately by a dashed line.

The lowest dashed line marks the contribution of the background estimated from the curvature of the pre-edge background. The line above stands for the contribution of the pre-edge background plus the Y signal. The weighted sum of reference spectra superposing the experimental curve coincides well with the experimental spectrum data despite slight differences in the shapes of the sharp white lines.

The graph below the original spectrum shows the differences between fit and spectrum data. The typical symmetrical shape of the residuum in the energy-loss ranges of the white lines arises from the difference in width and height of the white-line peaks. Nevertheless, the residual integral intensities within the energy-loss ranges used for quantification averages out, which results in negligible residual intensity contributions on the scale of the accuracy of this method for both integral intensities of Y and Zr.

Element-specific integral intensities $I_{Zr\text{-}L_{2,3}}$ and $I_{Y\text{-}L_{2,3}}$ within wide energy-loss windows of in each case 125 eV from edge onset (marked as shaded areas in Fig. 3.22) were used for the quantification. These were determined by multiplying the element-specific MLLS-fit parameter with the integral intensity of the respective reference spectrum in the specific energy-loss range.

Both extracted integral net intensities of Y and Zr, respectively, along this line scan are shown in Fig. 3.24a. As expected from single spectra of 8YDZ (cf. Fig. 3.23), the intensity of Zr is about five times higher than the net intensity the Y-L ionization edge. This is mainly attributed to the similarity of the shapes of the ionization edges as well as the chemical composition of 8YDZ with 17 mol% Y_2O_3.

Minor variations of both signals, which in principle run parallel, correlate with the variations of the MLLS fit parameter for the background intensity (compare Fig. 3.24a with 3.24b).

To verify the quality of the background MLLS fit parameter as measure for the local thickness, the normalized integral pre-edge background intensity in the energy-loss range of 1960–2060 eV is plotted in Fig. 3.24b. Since both graphs are almost equal, the fit parameter excellently represents the background intensity.

Since the pre-edge background far from ionization edges at lower energy losses is assumed to be directly thickness depending, it is concluded that the variations of the element-specific net signals can be mainly explained by the change of local sample thickness.

For the direct quantification of each spectrum of the line scans of Fig. 3.22, the factor $\hat{k}_{Zr,Y} = 0.99 \pm 0.02$ was determined by averaging the k-factors $k_{Zr,Y}$ determined from the sum spectra of each line scan. Therefore, Eq. 3.1 was used to correlate the measured intensities with the known chemical composition of the as-sintered 8YDZ, namely

$$\frac{C_{Zr}}{C_Y} = k_{Zr,Y} \cdot \frac{I_{Zr\text{-}L_{2,3}}}{I_{Y\text{-}L_{2,3}}} \tag{3.1}$$

Using this factor $\hat{k}_{Zr,Y}$, the local Y content was determined as described by Eq. 3.2, which is analogous to the equation used for quantitative EDXS (cf. Eq. 2.6). Therefore, the sum of the cation concentrations has been assumed to amount to 100 %. Analogously, the Zr content was calculated from the net intensities for each spec-

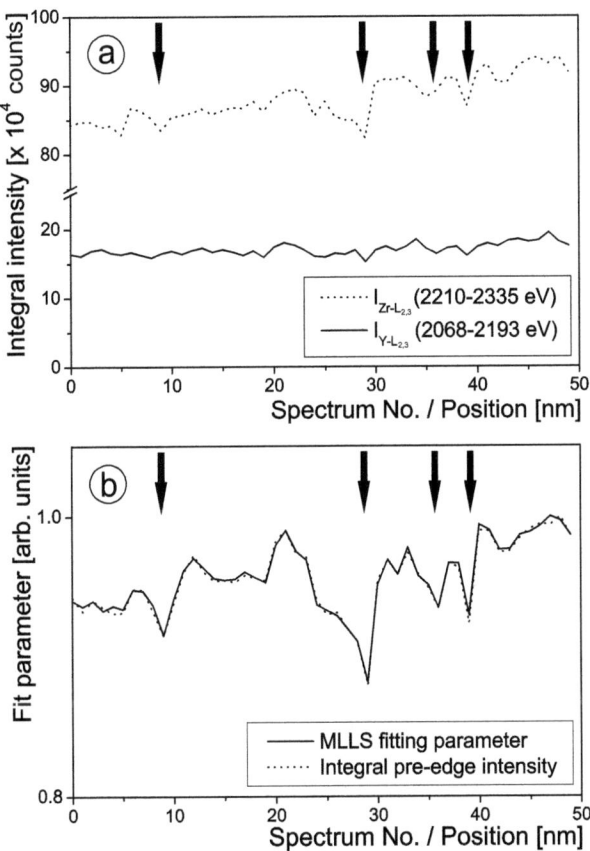

Figure 3.24: a) Integral intensity of Y and Zr derived from MLLS fitting, b) normalized MLLS parameter for the reference background in comparison with the normalized integral pre-edge background intensity (1960–2060 eV).

trum.

$$C_Y = \frac{I_Y}{\hat{k}_{Zr,Y} \cdot I_{Zr} + I_Y} \qquad (3.2)$$

Both the distributions of the Zr as well as the Y content along the line scan resulting from the net intensities (Fig. 3.24a) are shown in Fig. 3.25a. The values for the Zr and Y content vary in the ranges of 82–83.8 at% (83±0.5) and 16.2–18 at% (17±0.5), respectively.

The second graph (Fig. 3.25b) presents the thickness-corrected net intensities for both cations using the MLLS fit parameter for the background to correct the measured intensities of Fig. 3.25a. Therefore, each integral intensity of Y and Zr, respectively, was divided by the related fit parameter. Comparing the variations in the distributions of the thickness-corrected net signals with the directly quantified Zr and Y contents leads to the conclusion that the accuracy of the direct quantification is slightly better than the thickness-correction method. Nevertheless, this procedure of using a spectral energy range, which was recorded with the element-specific signal, yields the best choice if one likes to compare element-specific signals in regions where the thickness during/after the experiment is not known accurately, i.e., regions that became damaged during the measurement.

In summary, it can be stated that the ranges of variation of the local Y and Zr contents by quantitative EELS, which were similar for all studied line scans, are independent on the local sample thickness in the investigated range of sample thicknesses. The final accuracy is not influenced as much by statistics. Furthermore, it is mainly limited by the quality of the reference spectra, i.e., the shape of the reference spectra with respect to the experimental data, leading to finite residua between the measured data and the MLLS fit.

Oxygen Content The original intention of the simultaneous recording of the Y-$M_{4,5}$ and Zr-$M_{4,5}$ ionization edges and the O-K edge in energy-loss range II was the quantification of both the oxygen as well as the cationic contents along the line scans by the k-factor method.

However, since it was not possible to separate the net intensities of the M ionization edges of Y and Zr in a reliable manner, the local oxygen net intensity could not be quantified with respect to the cationic intensities. Nevertheless, the homogeneity of the oxygen content was conveniently evaluated by thickness-correcting the element-specific net signal. Therefore, the extracted and thickness-corrected net intensity of the O-K ionization edge was used as measure for the local oxygen content.

For signal integration, an energy window $\Delta = 50$ eV from the edge onset (532 eV) was used to minimize the influence of local properties (ELNES) on the evaluated integral intensity. To prevent artificial variations (cf. §2.2.3) of the oxygen net signal, which

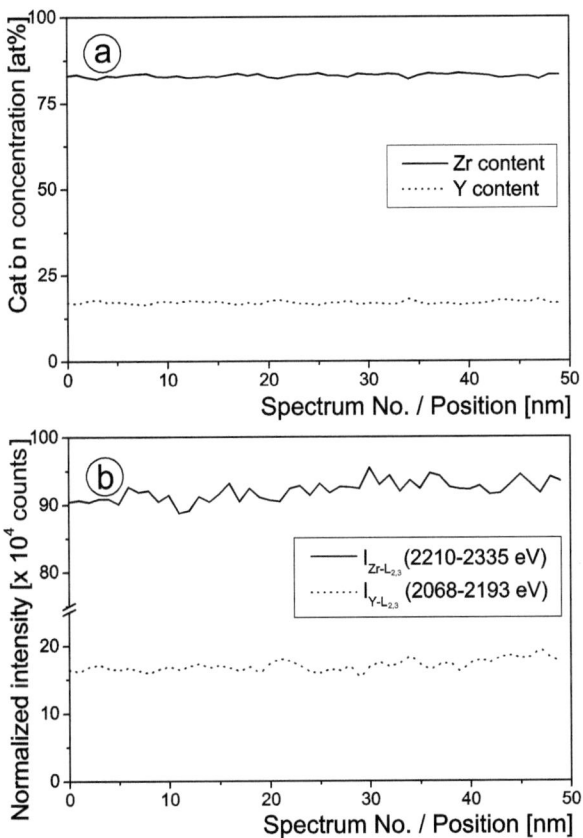

Figure 3.25: a) Quantified Y and Zr contents of 8YDZ-AS along the line scan presented in Fig. 3.24. b) Thickness-corrected net intensities of Y and Zr.

may arise from the well-established procedure by background modelling and subtraction for each individual spectrum along a line scan, MLLS fitting of the spectra of the line scans was utilized using *intrinsic* reference spectra.

Therefore, a background reference as well as the O-K edge reference were obtained from the sum spectrum of each line scan to evaluate this specific set of spectra. Multiplying the distribution of MLLS-fit parameters for the O-K reference spectrum with the integral intensity of the reference spectrum (532–582 eV) leads to the final distribution of the O-K net signal along each line scan. One representative distribution along a line scan is shown exemplarily in Fig. 3.26a. It exhibits similar untrustworthy variations as the net signals of Zr and Y in Fig. 3.24a.

For thickness correction, after normalization with respect to the highest value the integral intensity I_{BG} in the energy-loss range of 700–750 eV was used for thickness correction. The resulting distribution of the thickness-correction factor N along the line scan is plotted in Fig. 3.26a. To achieve thickness correction of the O-K net signal, each value along the line scan was divided by the related value N of this distribution.

The quality of this factor was tested by a second integral pre-edge background (135–160 eV) directly situated in front of the Y-$M_{4,5}$ and Zr-$M_{4,5}$ ionization edges. For comparison, the distribution of this value is plotted as dashed line in Fig. 3.26a.

The trends of both plots are similar, regardless of the different element-specific contributions in the evaluated energy-loss ranges. This leads to the conclusion that both integral intensities situated more or less far from element-specific ionization edges at lower energy losses are suitable as measure for the local sample thickness.

Finally, the thickness-corrected distribution of the O-K net intensity (Fig. 3.26a) is shown representatively in Fig. 3.26.

The mean value of I_{O-K} for this representative line scan arises to 450938 ± 4238 (1 σ) counts, resulting in a relative error of about 1 %. This accuracy counts for all evaluated line scans regardless of the local sample thickness within the investigated range of sample thickness, as discussed before for the evaluation of the L ionization edges.

The accuracy of this semi-quantitative treatment of the element-specific net intensity, resulting in slight variations of only one percent along the line scan (Fig. 2.11), is of similar order as the variations of the Y and Zr contents obtained by the direct quantification using a k-factor (Fig. 3.25) in spite of the necessity of the normalization with respect to the local sample thickness.

Despite high integral intensities in EELS compared to EDXS, the accuracy of the quantitative and semi-quantitative analyses is methodically limited to relative errors of a few percent. The most important reasons are summarized in the following. While for the quantification of the MLLS-fitted spectra, the quality of the reference spec-

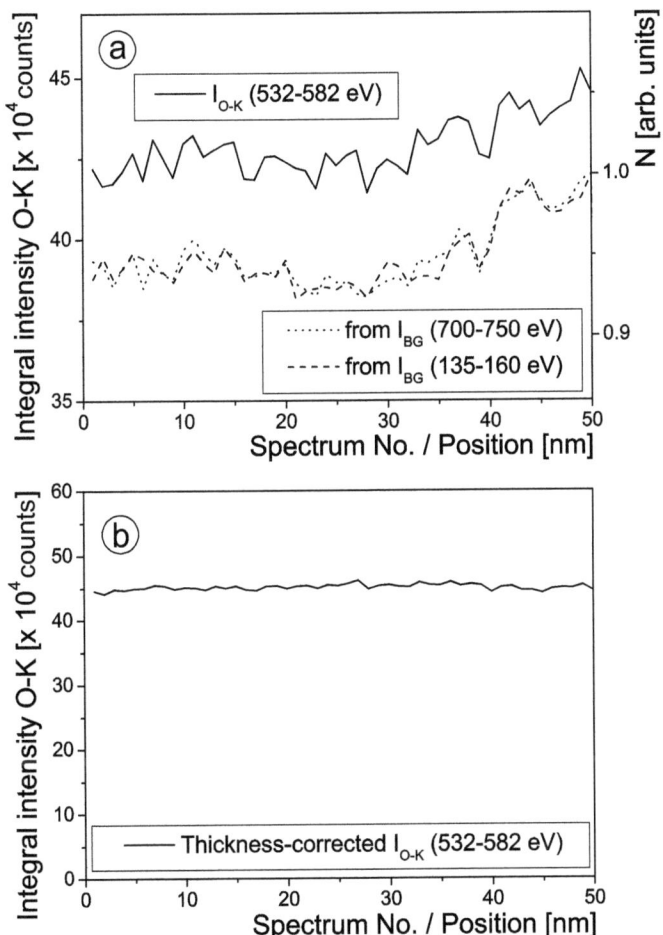

Figure 3.26: a) Integral net intensity of the O-K ionization edge ($\Delta = 50$ eV from edge onset) obtained from MLLS fitting (straight line). Factor for thickness normalization obtained from post-edge (700–750 eV) and pre-edge (135–160 eV) background. b) Thickness-corrected O-K net intensity.

tra has to be taken into consideration, the normalization with respect to the local thickness along the line scans plays a key role for the evaluation of an individual element-specific signal. First of all, the extraction of the net signals from the spectral data is limited by several factors.

1. Accuracy of the determined net signals is limited by

    - Quality of reference spectra (contour)
    - Slight energy drift of the spectrometer within a line scan
    - Residuum of the MLLS fitting is of the order of a few percent of the count rate
    - Quality of the background estimation

2. Accuracy of the factor for thickness correction is limited by

    - Integral used background intensity

Since the above-mentioned factors are of the order of the resulting variations of the Y, Zr, and O signals, any chemical inhomogeneity in the as-sintered 8YDZ has to be of the order of or smaller than the observed variations in EELS. Furthermore, one has to take into account the propagation of the primary electron beam through the TEM sample, the thickness of which is limited to a minimum of about 10 nm, and thus signal averaging.

Decomposition of 8YDZ at High Temperatures

The limits of the utilized analytical techniques were outlined in the previous part on the basis of the evaluation of spectroscopic data of the as-sintered 8YDZ-ASH. This material has been assumed to be optimal homogenized on atomic scale during the heat treatment in the pure cubic phase field.

This was previously verified within the limitations of the applied analytical techniques. In the following, the comparison between the element distributions of the cations as well as oxygen in as-sintered and heat-treated 8YDZ (8YDZ-H) is drawn. It will be shown that the coarsened regions (see Fig. 3.7), observed in heat-treated 8YDZ by dark-field imaging (Fig. 3.7), are strongly depleted in Y ions. This inhomogeneity on the cationic sublattice results in variations of the oxygen content in the material.

Despite the lower sensitivity for variations of the local cationic composition, as shown in the previous part, the Y and Zr contents in the following were obtained by quantitative EDXS, which is suitable to detect the variations in 8YDZ-H.

To elucidate the corresponding O distributions, the thickness-corrected net signals obtained from simultaneously recorded EEL spectra (energy-loss range II) along the line scans are plotted in the same graphs.

Fig. 3.27 juxtaposes representative line scans of as-sintered 8YDZ-ASH (Fig. 3.27a) and 8YDZ-H (Fig. 3.27b) obtained at different thicknesses between 8 nm and 30 nm (thickness given for each plot). TEM sample regions with thicknesses below 10 nm turned out to be inappropriate due to predominant electron-beam induced damage, which is expected to cause the preferential release of oxygen ions. This is visualized in the two upper graphs, where especially unreliably strong variations and trends of the oxygen net signal are recognizable. Since even small holes along such line scans were observed in HAADF STEM imaging immediately after the measurements, these line scans are only presented for the sake of completeness. The statistical limitation of EDXS at such small samples thicknesses in addition impedes the detection of chemical variations on the cationic sublattice (compare both graphs of 8YDZ-ASH and heat-treated 8YDZ-H).

In contrast, sample regions with thicknesses in the range of 15–30 nm are well suited to investigate the chemical variations in 8YDZ-H after heat treatment.

As discussed, the Y distributions and the O-K net signals of 8YDZ-ASH, which are also shown in Fig. 3.27, represent the limitations for the detection of chemical variations. The element distributions of 8YDZ-H on the other hand representatively show the features, which are detectable after the heat treatment. Despite the low sensitivity of EDXS, strong variations of the local Y contents are observed in both distributions, which surpass the statistical limitations more than 3σ.

To correlate the local decrease of the Y content with widths on the scale of 5-10 nm with microstructural features, the corresponding dark-field STEM image intensity along the line scans is shown below each graph. Since this annular dark-field STEM micrograph was exposed using an intermediate camera length of the microscope, the detected intensity of tetragonal additional reflections led to the distinct intensity variations in the resulting micrograph. Thus, coarsened, mainly tetragonal regions as observed in conventional TEM dark-field imaging (cf. Fig. 3.7) appear bright, while the mostly cubic matrix appears darker.

Comparing the trends of the Y content with the local image intensities leads to the conclusion that the coarsened regions in 8YDZ-H are strongly depleted in Y ions.

Since in dark-field imaging only one of the three variants of the tetragonal phase (c-axis can be aligned parallel to all three cubic main axes) can be visualized in a single image, regions without increased image intensity but decreased Y content were also found along the line scans. Such a region is exemplary marked by the circle in Fig. 3.27.

The inhomogeneity of the Y content in 8YDZ-H causes slight but significant varia-

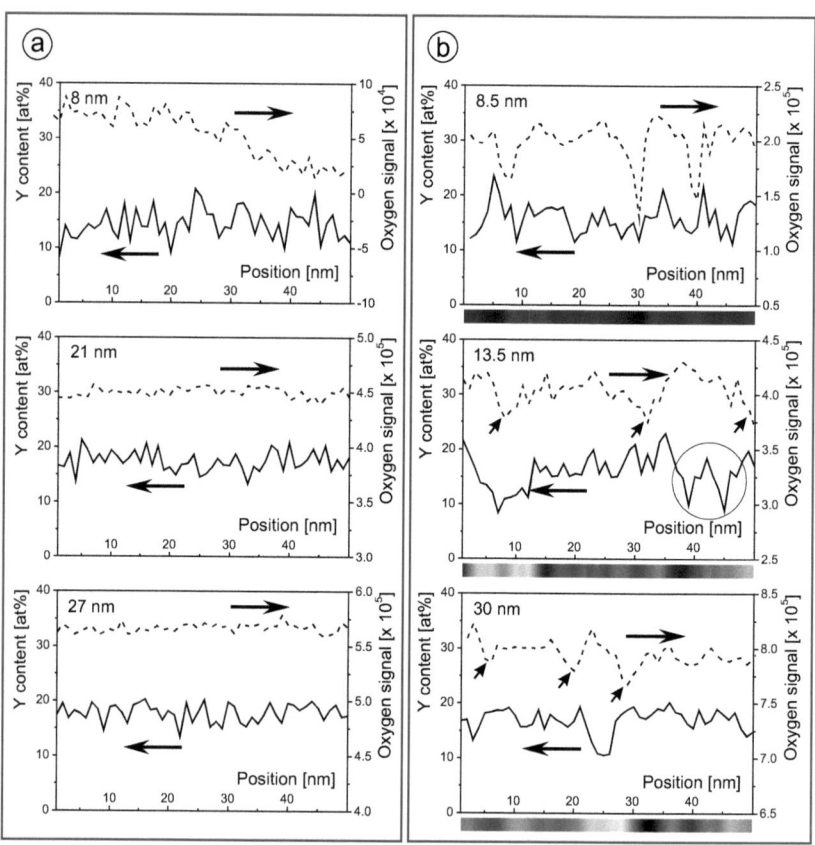

Figure 3.27: Y content (EDXS) with corresponding thickness-corrected O-K net intensity (EELS) of representative line scans at various thicknesses of a) 8YDZ-ASH and b) 8YDZ-H. The local thickness is given in each graph.

tions of the local O content. The corresponding O-K net signals are presented for each Y distribution.

At first appearance, both distributions run rather inversely in comparison to the plotted Y contents. From charge neutrality the decrease of the local oxygen-vacancy concentrations in the Y-depleted regions is expected due to the generation of oxygen vacancies by the presence of dopant cations (§ 2.1.1). This decrease of the local vacancy concentration results in the increased O-K intensity since this intensity is directly proportional to the number of oxygen ions in the probed sample volume.

But at second view, local drops of the O-K net signal below the mean value in the vicinity of the Y-depleted regions become recognizable. Such regions are marked by small arrows in both graphs. This signal decrease may indicate the tendency of the oxygen vacancies to accumulate next to the Y-depleted regions.

3.2.5 Raman Spectroscopy

Raman spectroscopy as contrary technique was applied to study the formation and evolution of the tetragonal phase in the 8YDZ thick-film electrolytes before (8YDZ-AS, 8YDZ-ASH) and after heat treatment (8YDZ-H). The reference spectra obtained from the 16YDZ single crystal and the polycrystalline 3YDZ specimen are presented in Fig. 3.28b. While 16YDZ (c-YDZ) only shows one pronounced maximum at a Raman shift of $603\,\text{cm}^{-1}$, the tetragonal 3YDZ specimen exhibits typical maxima [13, 148] at $262\,\text{cm}^{-1}$, $324\,\text{cm}^{-1}$, $466\,\text{cm}^{-1}$, and $646\,\text{cm}^{-1}$, respectively. Hence, from Raman spectra the crystal structures of both specimens can clearly be distinguished by this technique.

Two representative spectra of 8YDZ-AS and 8YDZ-H are shown in Fig. 3.28a. Vibrational modes due to tetragonal symmetry were excited in the heat-treated specimen (8YDZ-H). Thus, the spectrum of 8YDZ-H clearly shows maxima at wave numbers of $257\,\text{cm}^{-1}$, $327\,\text{cm}^{-1}$, and $468\,\text{cm}^{-1}$ corresponding to the above-mentioned of 3YDZ. The differences in the wave numbers may arise from the differences of the crystal structures of the measured tetragonal-type phases. The second reason may be differences of the lattice parameters, which depend on the composition (cf. Fig. 2.5).

Any excitation of tetragonal-type Raman bands in the spectra of 8YDZ-AS (Fig. 3.28a) was not identified despite the fact that tetragonal-type nanoscaled precipitates were found in this specimen by TEM and SAED (cf. § 3.2.2). Raman spectroscopy failed for the YDZ thin-film electrolytes since the intensity contributions due to the YDZ thin films with respect to the background of the sapphire substrate was too low.

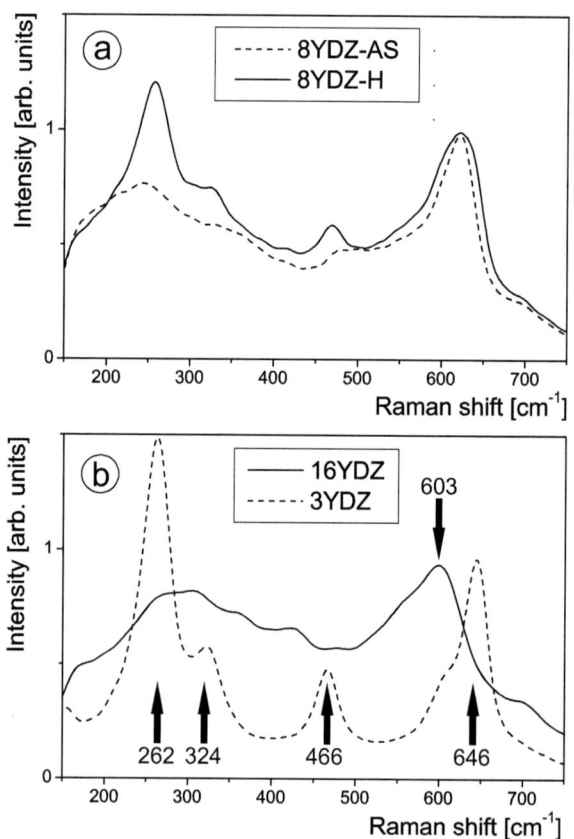

Figure 3.28: a) Raman spectra (Stokes scattering) of 8YDZ-AS and 8YDZ-H.
b) Raman reference spectra (Stokes scattering) obtained from 3YDZ (t'-YDZ) and 16YDZ (c-YDZ). All spectra are normalized with respect to the Stokes maximum at the highest wave number.

3.3 Nanocrystalline YDZ Thin Films

The following part describes the properties of the YDZ thin films, which were prepared in order to study the influence of grain size and thus grain-boundary density on ionic conductivity in YDZ. The detailed microstructural and chemical characterization is indispensable to reliably evaluate and discuss the findings of the electrical characterization that was performed by C. Peters from the *Institute of Materials for Electrical and Electronic Engineering* in the context of his PhD thesis [16]. His results are resumed in § 2.1.3.

Microstructural aspects like film thickness, grain size, and the distribution of the observed phases were studied by bright-field as well as dark-field TEM techniques in combination with electron diffraction. HAADF STEM imaging was utilized to visualize the distribution of pores in the thin films, which exhibit remarkable residual porosity (specimens YDZ-650, YDZ-850, and YDZ-1000), since this method is suitable to visualize local density variations. HRTEM in combination with EDXS was utilized to investigate grain-boundary regions with respect to the segregation of impurities during the calcination.

As mentioned in § 3.1.2, annealing temperatures up to 1350 °C were used to adjust the mean grain size from the nano- to the microscale. The specimen heat-treated at 1600 °C (YDZ-1600) was used as reference for phase analysis since it can be expected to be optimal homogenized in the pure cubic phase field of the phase diagram (see § 2.1.2).

3.3.1 Microstructure

Thin-Film Quality

SEM was carried out to show the high quality of the prepared YDZ thin films on the sapphire substrates. Representative top-view and cross-section images (imaging using secondary electrons) of the specimens are shown in Fig. 3.29 for all calcination temperatures. All films are free of cracks on the whole field of view of minimum 40 x 50 mm². The cross-section analysis exhibits film thicknesses between 390 and 450 nm. The adhesion to the sapphire substrates is excellent. No indication for the delamination of the films was found, which might occur during the calcination of the thin films due to different thermal expansion coefficients of YDZ and the sapphire substrates. Heat treating such films at 1600 °C (YDZ-1600) results in the disruption of the YDZ thin film (as shown in Fig. 3.30) where single grains reached lateral dimensions of more than 20 µm.

To gain information about the morphology and chemistry especially of the nanocrystalline specimens, SEM is limited not only by a worse lateral resolution in comparison

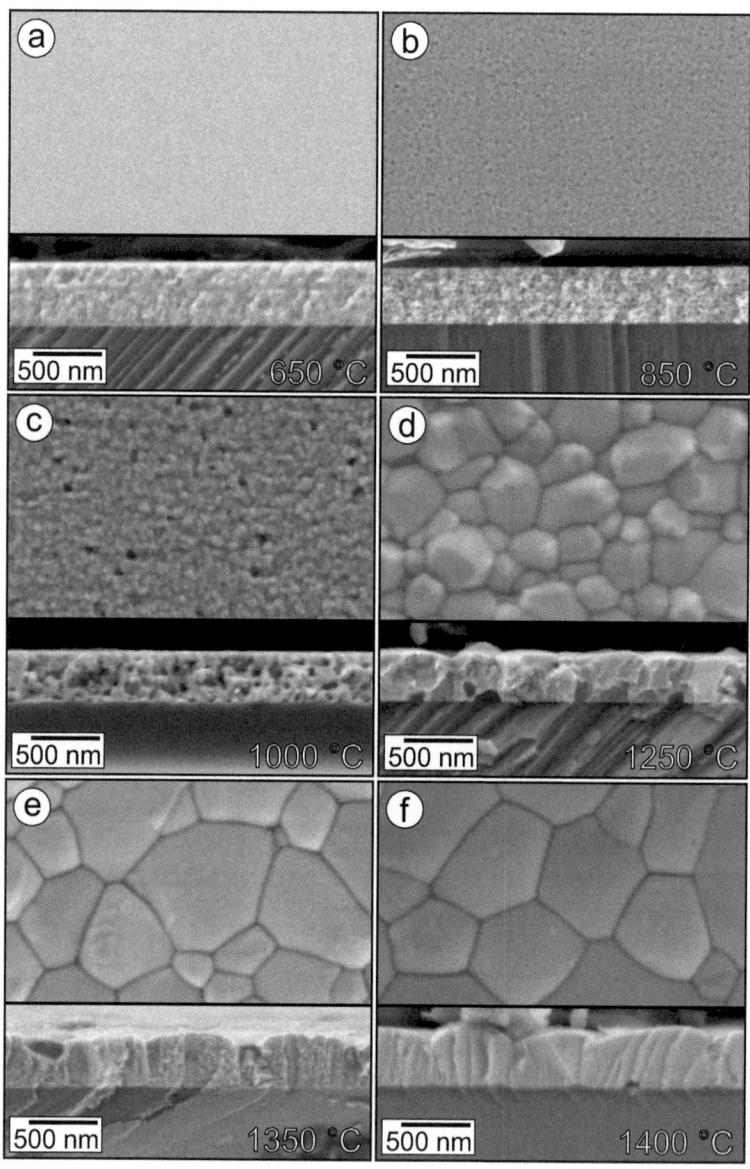

Figure 3.29: SEM top-view and cross-section images of a) YDZ-650, b) YDZ-850, c) YDZ-1000, d) YDZ-1250, e) YDZ-1350, and f) YDZ-1400.

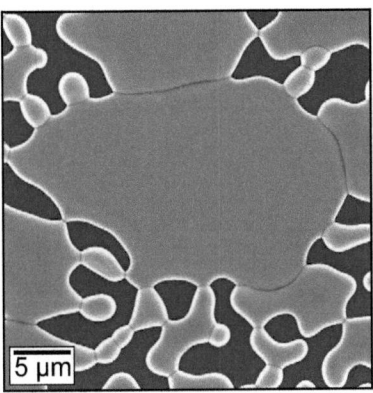

Figure 3.30: Top-view SEM image of YDZ-1600.

to TEM, but also by signal averaging in the bulky sample. This methodological limitation is due to the size of the interaction volume of the primary electrons that is much larger in SEM than in TEM. As example for this limitation, the porosity in the nanocrystalline specimens YDZ-650, YDZ-850, and YDZ-1000 is mentioned. The pores can only be vaguely recognized in the SEM images of these specimens (cf. Fig. 3.29a-c). Hence, various TEM techniques were applied in order to gain additional information about the micro- and nanostructure of the thin films. The experimental findings are summarized in the following.

Morphology

Fig. 3.31 shows TEM bright-field cross-section images of the thin films after the final heat treatment (YDZ-650 – YDZ-1350) to give an overview of grain size, film thickness and microstructural homogeneity of the prepared films.

The specimens YDZ-650 – YDZ-1000 in good approximation consist of isotropically grown grains as can be seen in Fig. 3.31a–c, while the size of the grains reaches the film thickness in YDZ-1250 and YDZ-1350. For YDZ-1350, the mono-layer type film consisting of brick-shaped grains is depicted in Fig. 3.31d.

Specimen YDZ-650 (Fig. 3.31a) obviously shows a layered morphology that is explained by the 10-fold dip coating. In contrast, a layered microstructure was not detected by bright-field imaging in the specimens, which were annealed at temperatures exceeding 650 °C. In those cases, a layer-type morphology is not observed in the micrographs of Fig. 3.31b-d. As will be shown later, the specimen heat-treated at 850 °C (Fig. 3.31b) also exhibits residual density variations due the 10-fold dip coating.

The specimens heat-treated up to 1000 °C exhibit nearly uniform film thicknesses

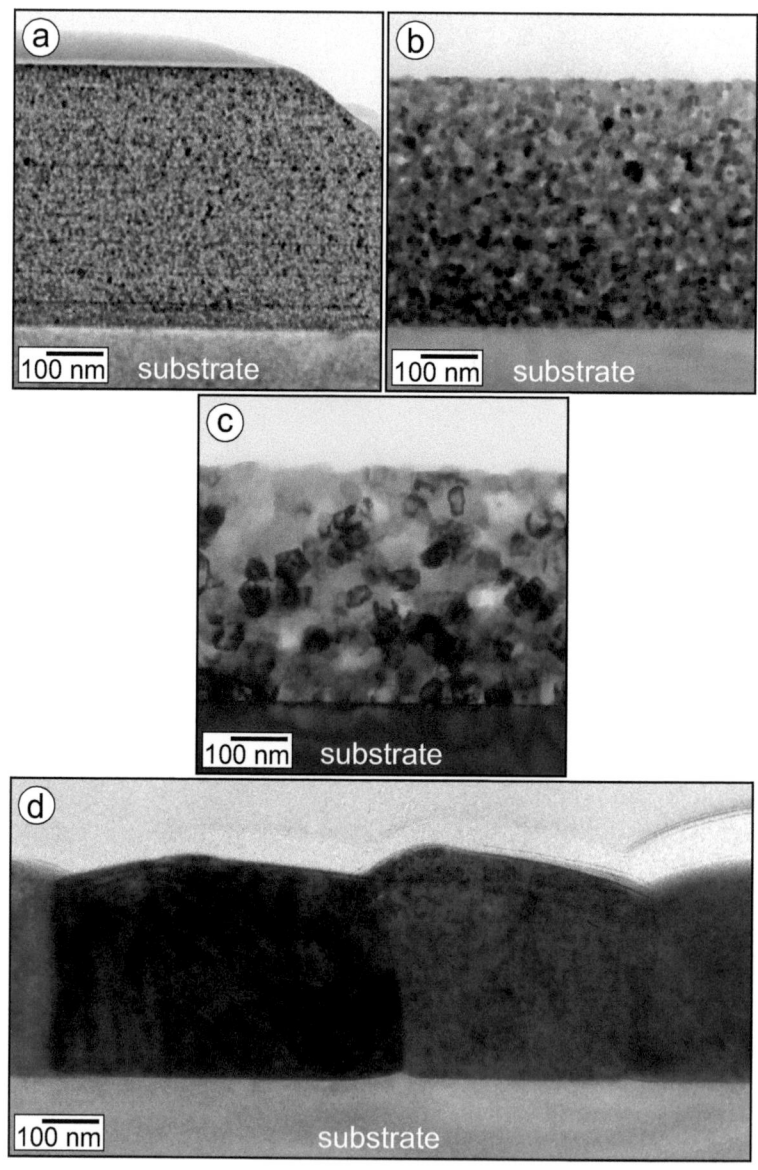

Figure 3.31: TEM cross-section bright-field images of a) YDZ-650, b) YDZ-850, c) YDZ-1000, and d) YDZ-1350.

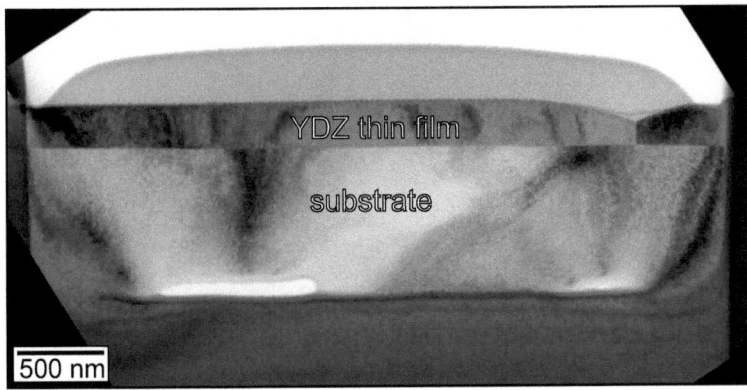

Figure 3.32: TEM cross-section bright-field image of a FIB lamella of YDZ-1600. The field of view shows one of the electron transparent window of the lamella consisting of two grains and a single grain boundary on the right-hand side.

(Fig. 3.31a–c), whereas the local thickness of the thin films of YDZ-1250 and YDZ-1350 (Fig. 3.31d) varies due the large lateral extension of the grains, reaching or even exceeding the film thickness. Nevertheless, apart from being uniform in thickness for one specific tempering temperature, the averaged electrolyte film thickness decreases with increasing annealing temperature for the specimens heat-treated up to 1250 °C as summarized in Table 3.3. This is due to the remaining volume fraction of porosity in those films, as will be shown later. This volume fraction was found to decrease with increasing calcination temperature. The thickness increase induced by porosity in relation to the thickness of the dense, pore-free films is used to estimate the volume fraction of pores as given in Table 3.3. For all annealing temperatures, the same density of the crystalline phase is assumed. Thus, the porosity decreases from about 15 vol% in YDZ-650 to zero in the dense films, as indicated in the previous work [189].

The contamination of the initially prepared YDZ-1600 (Fig. 3.32) specimen with Si during the heat treatment at 1600 °C necessitated the re-preparation of this specimen. The second specimen was only dip-coated 7 times. Consequently, the final thickness of 270 nm is slightly lower than the thickness of the other thin-film specimens.

The estimated size of grains in YDZ-650 was measured on the basis of HRTEM images, since conventional bright-field images could not be evaluated reliably due to the small size of the grains with respect to the TEM sample thickness. Grain-size distributions for the specimens YDZ-850 – YDZ-1350 were determined by the analysis of plan-view bright-field TEM images. Therefore at least 200–300 different grains were evaluated by the following procedure. The grain-boundary distributions of those

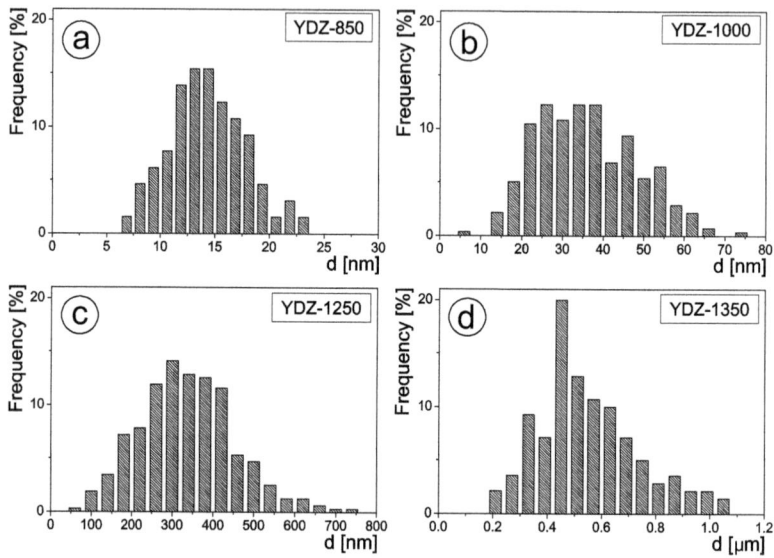

Figure 3.33: Grain-size distributions of a) YDZ-850, b) YDZ-1000, c) YDZ-1250, and d) YDZ-1350. For YDZ-1350 the lateral size of the grains is given in d).

bright-field images were determined using automated image-processing techniques. Next, the grain boundaries were displayed as threshold images. The projected grain area A of individual grains within those threshold images was approximated by a circle with the diameter $d = 2r = 2\sqrt{A/\pi}$.

Fig. 3.33 shows the resulting grain-size distributions of YDZ-850, YDZ-1000, YDZ-1250, and YDZ-1350. For better comparability, the distributions were normalized with respect to the number of evaluated grains. All specimens exhibit normal grain growth, as can be concluded from the unimodally distributed grain sizes (only one defined maximum in each grain-size distribution), independent of the final annealing temperature.

Both, the mean grain size for each specimen as well as the standard deviation, which are summarized in Table 3.3, were determined from the distributions in Fig. 3.33. Both values strongly increase from the nanoscale to the microscale with increasing annealing temperature. For YSZ-1250 and YDZ-1350, the average grain size corresponds to the lateral extension of the brick-shaped grains (Fig. 3.31).

Specimen	Thickness [nm]	Mean grain size d [nm]	Porosity [vol%]
YDZ-650	450	5	15.4
YDZ-850	420	14.2 ± 3.6	7.7
YDZ-1000	400	35.9 ± 12.3	2.6
YDZ-1250	390	334.9 ± 116.6	-
YDZ-1350	390	547.6 ± 186.9	-

Table 3.3: Microstructural characteristics of the YDZ thin films.

Porosity

A further important point of interest is the visualization of any remaining porosity, which might be present due to the fabrication procedure. Since TEM bright-field as well as dark-field imaging are inappropriate due to the strong dynamical effects, depending on local thickness and orientation, HAADF STEM imaging was utilized to visualize local changes of the density due to porous regions. Since porous regions do not scatter any primary electrons in comparison to the dense YDZ grains, pores will appear darker in comparison to the dense ceramic regions. The sapphire substrate appears much darker in comparison to the YDZ due to the lighter ions on the cationic sublattice.

Representative dark-field STEM images of the specimens with significant porosity, i.e., YDZ-650, YDZ-850, and YDZ-1000, are presented in Fig. 3.34. In all three specimens pores with dimensions similar to the mean grain size are found. This becomes obvious if one compares the HAADF STEM images of Fig. 3.34 with the corresponding bright-field images presented in Fig. 3.31 (same magnification). The YDZ thin films annealed at temperatures exceeding 1000 °C were found to be free of pores with the exception of very few and tiny interstices between several grains observed in bright-field imaging.

The layered morphology of YDZ-650, which was observed in bright-field cross-section images (Fig. 3.31a), is also recognizable in the STEM image (Fig. 3.34a) of this specimen. From this STEM image, it can be concluded that the morphology is characterized by variations of the density of the thin film parallel to the surface of the film. Equidistant layers of higher density than the average are followed by a more porous layer. This is indicated by a slight increase of image intensity as shown in the plot in Fig. 3.34a. This plot visualizes the averaged image intensity parallel to the interface substrate/YDZ thin film.

It has to be noted that not all 10 sublayers are present in the field of view of Fig. 3.34a. The uppermost two sublayers of the thin film were etched away during Ar^+-ion etching in the displayed region of the TEM sample. The dashed line represents the initial surface of the thin film. The slight decrease of the overall intensity is explained by the wedge shape and thus the local thickness of the TEM sample (standard preparation). In contrast to the bright-field image (Fig. 3.31b), the HAADF STEM image of YDZ-850 (Fig. 3.34b) also demonstrates such density variations due to the 10-fold dip-coating procedure. This is visualized by the overlaid plot of the averaged image intensity, which reveals 10 nearly equidistant maxima indicating density variations parallel to the thin film/substrate interface.

Since the grain size and thus pore size in YDZ-1000 reaches the thickness of the individual sublayers, the layered morphology is not present anymore in this specimen

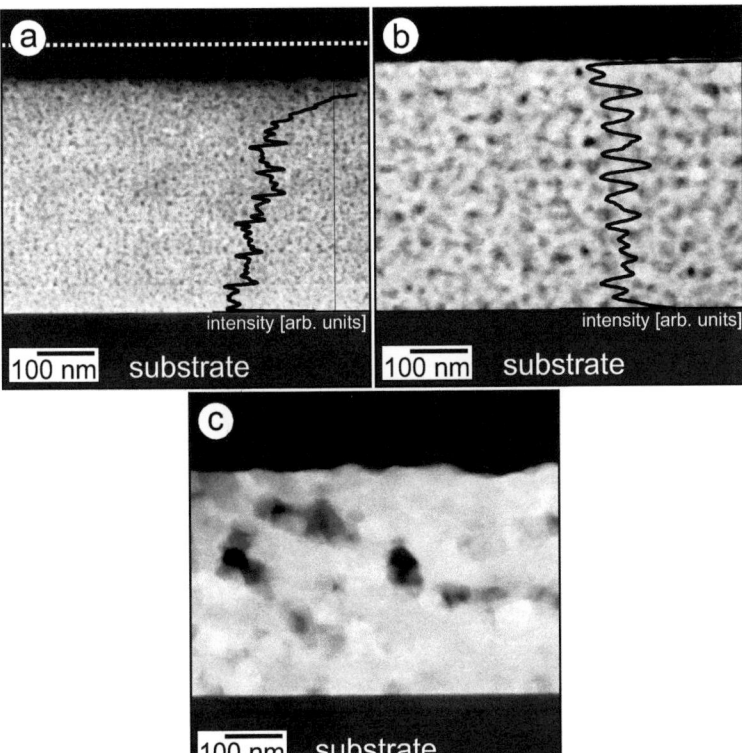

Figure 3.34: HAADF STEM cross-section images of a) YDZ-650, b) YDZ-850, and c) YDZ-1000. In a) and b) the averaged (parallel to the substrate/thin film interface) image intensity is overlaid to visualize the intensity variations in the thin films explained by variations of the local density due to the 10-fold dip-coating process.

(cf. Fig. 3.34c). It has to be mentioned that both, YDZ-850 and YDZ-1000 were prepared by the FIB lift-out preparation technique, resulting in plane-parallel TEM samples with more homogeneous thickness in comparison to the YDZ-650 TEM sample. Hence, the mean image intensity of the YDZ is nearly constant in both STEM images.

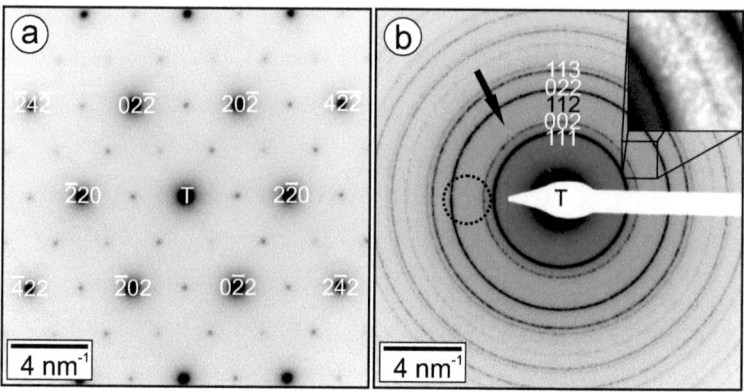

Figure 3.35: a) SAED patterns of a single grain of YDZ-1350 aligned along the [111] ZA. b) Debye SAED pattern of YDZ-1000: the faint ring of additional {112} reflections marked by the arrow in b) is presented in the insert with enhanced contrast. The dotted ring corresponds to the smallest objective aperture that was used for dark-field TEM imaging (see next part).

3.3.2 Structure

Crystal Structure

To gain information about the phases present in the sol-gel derived YDZ thin films, SAED was performed in addition to XRD as outlined in § 2.2.1. To record single-crystallite SAED patterns of YDZ-1250 and YDZ-1350 along specific directions the smallest SA aperture with a diameter corresponding to about 220 nm was used. This procedure yields the possibility to directly visualize double-diffracted reflections in thin TEM sample regions and thus to clearly identify tetragonal symmetry in the material as discussed by Butz and co-workers [15]. In contrast, larger SA apertures in combination with plan-view TEM samples had to be used to obtain Debye diffraction patterns of the specimens with grain sizes smaller than the smallest SA aperture, i.e., YDZ-650, YDZ-850, and YDZ-1000.

Fig. 3.35a shows a SAED pattern obtained from a single grain of YDZ-1350 (plan-view TEM sample) along the [111] zone axis. In accordance with the thick-film electrolytes (cf. Fig. 3.5) the observed reflections are induced by the presence of the cubic and a tetragonal-type phase if one assumes three variants with the c-axis aligned parallel to all three cubic ⟨100⟩ axes. For simplicity the cubic indexing scheme is used in Fig. 3.35. Further SAED patterns of this sample along the [114] and [233] zone axes (not shown here) do not yield any indication for the presence of the monoclinic phase. The same findings hold for YDZ-1250. As outlined in more detail in Ref. [15], the inner ring of {110} reflections in Fig. 3.35, which are kinematically forbidden both for the cubic and the tetragonal phase, disappears in very thin TEM sample regions indicating double-diffracted reflections.

Debye electron diffraction patterns were recorded for the specimens annealed at temperatures T≤1000 °C. Representative for these specimens, a Debye pattern of YDZ-1000 is shown in Fig. 3.35b. This diffraction pattern consists of homogeneous rings of reflections due to the large number of simultaneously illuminated nano-scaled grains which were selected by the SAED aperture. The strongly excited rings of reflections marked by white indices are, as in Fig. 3.35a, mainly induced by the cubic phase. Nevertheless, a faint ring of {112} reflections, which is the most pronounced type of additional reflections by any tetragonal phase, can be barely recognized. For better visualization, this ring (marked by the arrow in Fig. 3.35) is shown enlarged and contrast enhanced in the insert.

Any splitting of reflections that would indicate different lattice parameters for the cubic and the tetragonal-type phase can not be resolved within the accuracy of this technique.

SAED studies of specimen YDZ-1600 are of particular interest because this thin film was heat-treated at 1600 °C in the pure cubic phase field according to the published data. However, reflections due to tetragonal symmetry are also clearly visible in single-crystallite SAED patterns of YDZ-1600.

Distribution of the Tetragonal Phase

The distribution of the tetragonal phase is visualized by dark-field imaging using kinematically allowed {112} intensity. Small precipitates of the tetragonal phase are observed in all films (Fig. 3.36) as for microcrystalline 8YDZ and 10YDZ (cf. §3.2.2). For YDZ-650, YDZ-850, and YDZ-1000 with nanoscaled grains, the smallest available objective aperture was positioned on the ring of {112} reflections as indicated by the dotted circle in Fig. 3.35b. Despite the small aperture size, contributions of the rings of {022} and {002} reflections could not be completely excluded, resulting in localised bright contrast due to the cubic matrix of specific grains. Whereas dark-field images of single grains of YDZ-1250 and YDZ-1350 were obtained by aligning single grains close to $\vec{g}=\{224\}$ two-beam condition with strongly excited {112} reflections.

The presence of the cubic phase in the dark-field images using {112} reflections is shown in Fig. 3.36a and b for YDZ-650 and YDZ-850 exhibiting the smallest grains. The bright regions, marked by arrows in both images, with sizes of about 5 and 15 nm, respectively, are due to the cubic matrix of grains with strongly excited {002} or {022} reflections. These regions show high intensity due to the high structure factors of the {002} and {022} reflections.

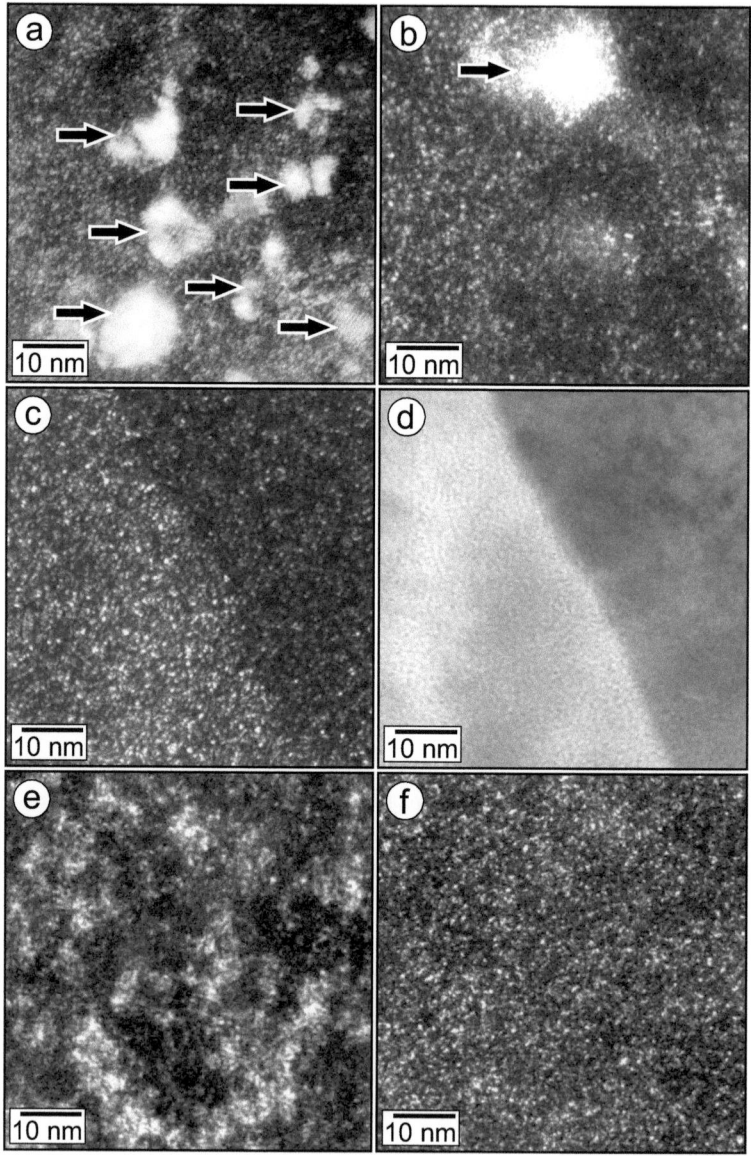

Figure 3.36: Dark-field TEM images of a) YDZ-650, b) YDZ-850, c) YDZ-1250, e) YDZ-1350, and f) YDZ-1600. d) Bright-field TEM image of the same region of YDZ-1250 as shown in c) to visualize the grain boundary in the field of view. The arrows in a) and b) mark contrast contributions due to reflections of the cubic phase indicating the cubic matrix of single grains.

Furthermore, in all other regions of the two images homogeneously distributed inclusions with intermediate contrast (within the dark cubic matrix of the grains) are found. They are due to tetragonal-type precipitates with sizes of a few Å. The change of contrast of the inclusions in different regions of both images may be related to different orientations of the embedding grains and thus to different local excitation of the $\{112\}$ reflections of the tetragonal phase. The continuous increase of the density of precipitates from the upper right to the lower left corner is caused by the increasing sample thickness and thus increasing number of projected precipitates.

Centred $\{112\}$ dark-field images of single grains were obtained for the specimens with microscaled grains, i.e., YDZ-1250, YDZ-1350, and the reference specimen YDZ-1600, by orienting the grains close to a $\langle 110 \rangle$ ZA ($\vec{g} = \{224\}$ two-beam condition). Fig. 3.36c, containing a grain boundary, shows the distribution of tetragonal precipitates in YDZ-1250. The grain boundary is for better visibility depicted in the corresponding bright-field TEM micrograph (Fig. 3.36d). While the lower left grain was ideally oriented for dark-field imaging (high contrast in Fig. 3.36c), the upper right grain shows lower contrast due to the misalignment. Nevertheless, contributing $\{112\}$ reflections could be selected by the objective aperture for both grains. Fig. 3.36c demonstrates the homogeneous distribution of the tetragonal precipitates. The density of tetragonal inclusions inside the grains and close to the grains boundary is homogeneous. This finding also applies to all other specimens.

Due to the large grain sizes in YDZ-1350 and YDZ-1600, the bright regions in Fig. 3.36e and f show the distributions of tetragonal precipitates within single grains. In contrast to all other thin-film specimens coarsened tetragonal precipitates are observed in YDZ-1350 as visualized by the inhomogeneous distribution of the bright regions.

Despite the high annealing temperature in the pure cubic phase region, sample YDZ-1600 also shows nanoscaled inclusions of the tetragonal phase that are homogeneously distributed, as it was found for the specimens YDZ-650 – YDZ-1250.

Furthermore, the increase of the density of tetragonal-type precipitates was found in YDZ-1600 in the vicinity of grain boundaries as well as at the interface between film and substrate. This is shown in the dark-field images in Fig. 3.37. The field of view, depicted in the images, shows the region close to the grain boundary that is visible in the bright-field image in Fig. 3.32. Both grains could be aligned optimal for the imaging of the tetragonal phase using a $\{112\}$ reflection. The cloudy contrast of the substrate in the lower dark-field image is due to damage induced by the Ga-ion milling during FIB sample preparation. Such distinct contrast was not found in conventionally prepared samples.

In summary, it was found, that tetragonal nanoscaled precipitates are homogeneously distributed in most of the specimens (YDZ-650 – YDZ-1250 and YDZ-1600) indepen-

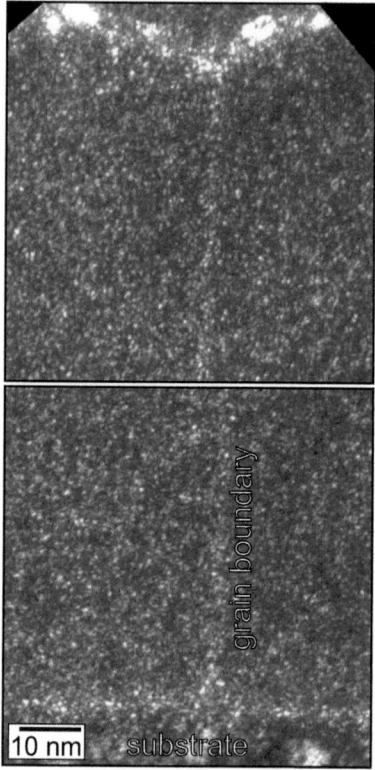

Figure 3.37: Dark-field images of the FIB lamella of YDZ-1600 (cf. Fig. 3.32) showing the distribution of tetragonal-type precipitates in the vicinity of the grain boundary and at the interface to the substrate. The lower micrograph shows the region at the interface between YDZ thin film and sapphire substrate. The upper image shows the surface-near region.

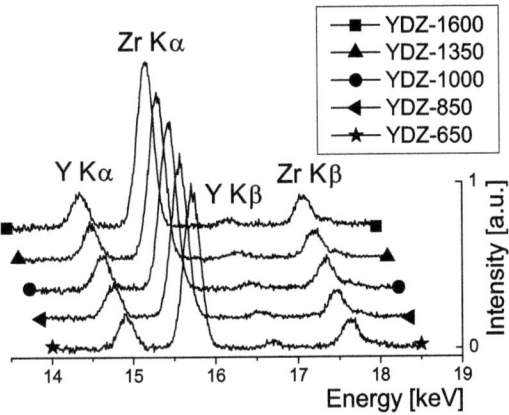

Figure 3.38: Representative SEM EDX spectra of large areas of the YDZ thin films in the energetic range of 14–18.5 keV depending on the heat treatment (primary electron energy 30 keV).

dent of grain core or grain-boundary regions. These precipitates tend to agglomerate in YDZ-1350. An increased density of tetragonal precipitates was found in YDZ-1600 in the vicinity to grain boundaries as well as at the interface film/substrate.

3.3.3 Chemistry

Composition of the Sol-Gel YDZ Thin Films

According to the sol-gel based process that was utilized to prepare the YDZ thin films, the dopant distribution on the cationic sublattice was assumed to be homogeneous. To determine the mean ratio of Y^{3+} and Zr^{4+} ions and thus the stoichiometry of the thin films after each calcination, quantitative EDXS analyses using a scanning electron microscope (primary energy of electrons 30 keV) were performed by illuminating large sample regions for recording the EDX spectra. This procedure is expected to yield minimized orientation effects on the obtained spectra. Representative background-subtracted spectra of the different specimens in the energy range of 14–18.5 keV are presented in Fig. 3.38. This spectral range contains both the Y and Zr $K\alpha$ and $K\beta$ emission lines as indicated. For better comparability, each spectrum was normalized with respect to the Zr $K\alpha$ peak intensity. The intensity ratios of the Y-K and Zr-K lines in the different spectra are nearly identical. It can therefore be concluded that the dopant concentration is similar in all specimens, independent of the temperature of the final heat treatment.

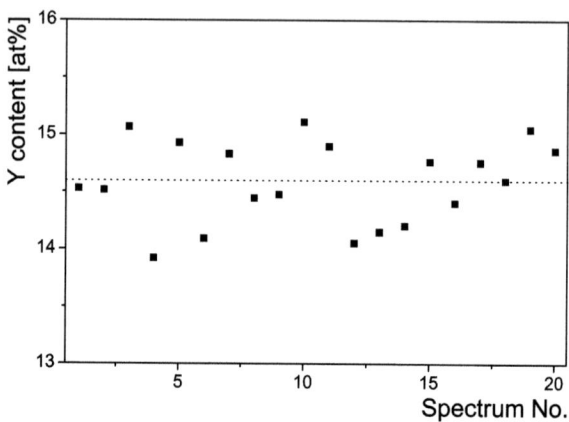

Figure 3.39: Y contents derived from 20 independent measurements of YDZ-1350. The calculated mean value of 14.6 at% Y is marked as dotted line.

First quantitative evaluations of these EDX spectra yielded a dopant concentration of 16.6 ± 0.6 at% Y^{3+} on the cationic sublattice in the YDZ thin films corresponding to 8.3 ± 0.3 mol% of Y_2O_3. This value was calculated by averaging the values for each thin film. This value is slightly higher than the value expected from the preparation of the sol-gel precursor by weighting out the basic materials, i.e., 14.75 at% Y.
More thorough quantification of 20 EDX spectra of a cross-section TEM sample of YDZ-1350, obtained in the CM200 electron microscope, was performed with respect to the as-sintered 8YDZ-AS thick-film electrolyte material with an given Y_2O_3 content of 8.52 mol%. Therefore, Eq. 2.6 (§ 3.2.4) was used with the determined k-factor of 1.02 (cf. § 3.2.4, page 68). The spectra were obtained from different sample regions by illuminating a few grains simultaneously. The resulting Y concentrations derived from these measurements are summarized in Fig. 3.39. The variations in this distribution are explained by statistical limitations due to the counts in the Y $K\alpha$ peak. The number of counts in this peak was about 3000–4000. Hence, the mean value of the quantified Y content of these 20 spectra arises to 14.6 ± 0.3 at% Y^{3+} (7.3 ± 0.2 mol% Y_2O_3) on the cationic sublattice. This value is in excellent agreement with the initially intended dopant concentration of 14.75 at% Y, i.e., 7.4 mol% Y_2O_3. The disagreement of the values obtained by analytical SEM and TEM may be due to geometrical aspects, i.e., the fact that the thickness of the thin films (about 400 nm) is too small for common quantification techniques (ϕ-ρ-Z method) with respect to the interaction volume of the primary electrons (30 kV) in SEM. A further reason may be secondary absorption and fluorescence, which may have had much more influence on the quantification of the K emission lines of the cations in analytical SEM in com-

parison to the TEM study.

Only for YDZ-1350, a local reduction of the Y content was found by local EDXS analyses with the CM200 electron microscope using a small probe size of about 1 nm. This finding is assumed to be due to inhomogeneities on the cationic sublattice as observed in heat-treated 8YDZ (8YDZ-H, cf. §3.2.4). The inhomogeneity of the distribution of the tetragonal phase in YDZ-1350 (cf. Fig. 3.36e) corroborates this assumption.

Scanning TEM in combination with electron energy-loss spectroscopy was applied to analyse possible reactions between the sapphire substrate and the YDZ thin film in the reference specimen, heat-treated at the highest temperature of 1600 °C (YDZ-1600). Line-scan experiments with a step size of 1 nm were carried out across the substrate/YDZ thin-film interface in order to investigate any interdiffusion at the interface. Using the Al-$L_{2,3}$ ionization edge, the Al-signal in the YDZ falls below the detection limit of the applied method in the distance of only a few nm from the interface, which excludes significant diffusion of Al into YDZ even at the highest annealing temperature. Despite the microstructural conspicuousness of the increased volume fraction of tetragonal precipitates at the grain boundaries as well as at the interface to the substrate, any Al could not been detected by analytical TEM techniques in the YDZ film of specimen YDZ-1600.

Grain Boundaries

The thorough characterization of the grain boundaries regarding the segregation of impurities is of particular relevance for the application of the nanocrystalline thin films in SOFC structures (cf. §2.1.2). HRTEM images of representative grain boundaries of nanocrystalline thin films, i.e., YDZ-850 and YDZ-1000, are shown in Fig. 3.40. The arrow-marked grain boundaries, representatively shown for all specimens, do not yield any evidence for amorphous grain-boundary phases. This even applies to YDZ-650, which was annealed at the lowest temperature, indicating the absence of unreleased carbon residues of the sol-gel precursor at grain boundaries within the limitation of the applied methods.

EDXS analyses using a fine electron probe with diameters of approximately 1 nm, in addition performed at numerous grain-boundary and triple-point regions, do also not indicate the segregation of impurities within the detection limit of 0.1 to 1 at% in the illuminated volume in any of the samples.

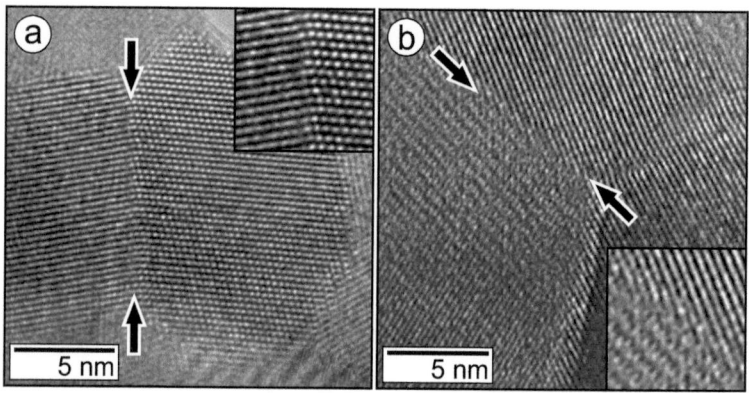

Figure 3.40: HRTEM images of a) YDZ-850 and b) YDZ-1000 showing oriented grain with representative grain boundaries.

Chapter 4

Discussion

4.1 Microstructural Evolution and Decomposition of 8YDZ

Numerous stable (t-, c-, m-YDZ) and metastable phases (t', t"-YDZ) in the Y_2O_3–ZrO_2 system doped with up to 10 mol% Y_2O_3 (cf. § 2.1.2) exist, which result in a large variety of possible microstructures, as outlined in § 2.1.2. This is summarized by Heuer et al. [5]. The experimental results obtained from the specimens in this study, i.e., microcrystalline 8.5 mol% YDZ, microcrystalline 10 mol% YDZ, and nano- as well as microcrystalline 7.3 mol% YDZ thin films, give new insights into the Y_2O_3–ZrO_2 phase diagram in the vicinity to the c — c+t phase boundary.

A tetragonal-type phase was found to homogeneously precipitate in nanoscaled regions in all as-prepared specimens irrespective of the preparation process and the dopant content (Fig. 3.6 and 3.36) [15, 35, 191, 192]. In the following, these precipitates will be denoted as TNP (tetragonal nanoscaled precipitate). Furthermore, the decrease of the volume fraction of these TNPs in microcrystalline YDZ is observed with increasing dopant content from 8.5 mol% to 10 mol% by SAED and dark-field TEM imaging [15, 191]. 16YDZ was found to be pure cubic except short-range ordering indicated in SAED [15, 191].

Distinct coarsening of the distribution of the TNPs was observed after heat treating microcrystalline 8YDZ thick-film electrolytes at temperatures of 950 °C (2500 h) and 1000 °C (1500 h, 3000 h, 5800 h), respectively. It has been also shown by analytical TEM that the microstructural evolution of 8YDZ is accompanied by the chemical decomposition of the material on the nanoscale. To some extent, a similar microstructural evolution was observed at higher temperature, i.e., 1350 °C, in the studied YDZ thin-film specimen YDZ-1350 after 24 h. Coarsening of the TNPs was not observed for 10YD at 950 °C after 1000 h [15].

It has to be added at this point that the YDZ specimens studied for this investigation, irrespective of the type of specimen, may not have reached thermodynamic equilibrium if this is defined as absolute minimum of the Gibbs free energy for a given composition and temperature. This has to be attributed to two facts. First, the diffusivity for cations in YDZ is rather slow (e.g. 5 nm in 1000 h at 1000 °C, according to Eq. 4.1 on p. 110) at temperatures below 1200 °C according to several authors [106–111]. Second, some specimens may have reached a local minimum of the Gibbs free energy function by the formation of metastable phases during cooling (cf. Fig. 4.1a). But it is still an open question whether this metastable state is the absolute minimum of the Gibbs free energy or not.

For the specimens rapidly cooled from the pure cubic phase field to low temperature (open symbols in Fig. 4.1a) it is concluded that the formation of the TNPs leads to such a local minimum of the Gibbs free energy. This seems to be the case independent on the composition in the studied dopant range, i.e., 7.3–10 mol% Y_2O_3. The material does not have the opportunity to reach the absolute minimum, i.e. the decomposed state for the YDZ thin films and 8YDZ, within finite time. This is attributed to the low mobilities of the cations, as mentioned before. No changes of the microstructure were observed for 8YDZ thick-film electrolytes, which had been stored in the laboratory at room temperature longer than 5 years.

The microstructure of 8YDZ continuously alters at around 1000 °C. Hence, the specimens 8YDZ-H, 8YDZ-L1500, 8YDZ-L3000 are not in thermodynamic equilibrium after the heat treatment. Even after 5800 h at 1000 °C (8YDZ-L5800), the material may not have reached equilibrium state.

Nevertheless, the experimental findings derived from dark-field TEM imaging in combination with the analytical results substantiate the expansion of the t+c two-phase field to higher Y_2O_3 contents, as schematically drawn in Fig. 4.1a. In this diagram a revised position for the c+t — c phase boundary towards higher Y_2O_3 contents is presented by the dotted line. Literature data [73] presented in §2.1.2 indicates that the boundary between the c+t and the pure cubic phase field may be given by the grey line in Fig. 4.1a (cf. Fig. 2.2). The new phase boundary was derived from experimental observations of specimens, which did not show any microstructural change (open symbols) and samples that clearly showed decomposition (filled symbols). Hence, the revised boundary line has to be situated between the open and the filled symbols. As discussed later, the grey symbols mark YDZ thin-film specimens, in which no significant microstructural changes have been found. Nevertheless, these specimens are expected to be situated in the c+t two phase field during preparation as well.

A revised position for the phase boundary of the metastable phase diagram between the pure cubic phase field and the c+t" two-phase field, as proposed by Yashima et al. [89], can not be deduced from the experimental results of this study. This is due to the fact that all specimens were characterized at room temperature. Furthermore, in-situ heating experiments in an transmission electron microscope in principle may show the disappearance of t"-YDZ at high temperatures. However, common heating holders for most transmission electron microscopes are limited to temperatures below 1100 °C. The covered temperature range appears to be insufficient to study this phase transition at high temperatures.

4.1.1 As-Sintered 8YDZ

Chemical Homogeneity

To prepare dense 8YDZ thick-film electrolytes, the green tapes were sintered at 1550 °C for 2 h [15, 166]. For comparison, a thick-film electrolyte (8YDZ-ASH) was subsequently homogenized in the pure-cubic phase field at 1700 °C for 4 h to assure a homogeneous dopant distribution on the cationic sublattice. Except for slightly larger grains in 8YDZ-ASH in comparison to 8YDZ-AS, due to the additional heat treatment, no structural and microstructural differences were found. From this fact it is concluded that, in combination with the used basic materials, the temperature of 1550 °C (2 h) is sufficient to chemically homogenize the investigated 8YDZ electrolyte foils.

The chemical homogeneity of an as-sintered 8YDZ specimen 8YDZ-AS was verified on the nanoscale by combined EDXS/EELS line-scan experiments within the limitations of both methods as shown in Fig. 3.25a and 3.26b. The final accuracy of the directly quantified Y concentration (cf. Fig. 3.25a) by evaluating the integral intensities of the Y-$L_{2,3}$ and Zr-$L_{2,3}$ ionization edges arises to $\pm 3\%$ of the absolute value (Fig. 3.25). Similar deviations were found for the thickness-corrected integral O-K signal as measure for the local oxygen content (cf. Fig. 3.26b). The quantification errors of EELS and EDXS were in detail outlined in § 3.2.4 and 3.2.4.

It was shown that the accuracies of the analytical techniques clearly allow the detection of compositional inhomogeneities in 8YDZ-H. Within the limits of EELS and EDXS, any lateral inhomogeneity of the Y concentration has to be smaller than a few percent of the given dopant value. However, signal averaging in the probed volume at finite sample thickness during the propagation of the primary electrons in addition limits the resolution.

Formation of t"-YDZ

Despite a dopant content of 8.5 mol% Y_2O_3, the 8YDZ-AS specimen is not completely stabilized in the cubic phase [15, 35]. Nanoscaled regions with tetragonal symmetry precipitate homogeneously in the material during cooling to room temperature as

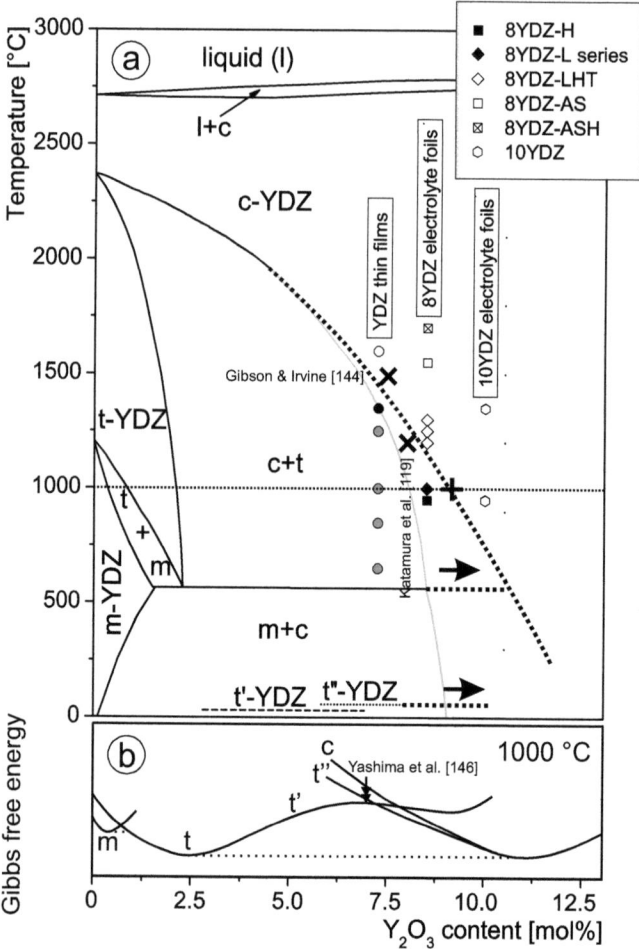

Figure 4.1: a) Revised Y_2O_3–ZrO_2 phase diagram in the zirconia-rich region (cf. Fig. 2.2). A new position of the c — c+t phase boundary at higher Y_2O_3 contents is proposed. The black cross at 1000 °C marks the solvus of Y in c-YDZ at 1000 °C, i.e., 9.2 mol% as calculated by the lever rule (Eq. 4.1.2 on p. 111) using the estimated volume fraction of Y-depleted regions in 8YDZ-L5800. The c+t ↔ c phase boundary, as derived from literature [73], is indicated by the grey line. Two solvuses proposed in literature [119, 144] are added. b) Proposed Gibbs free energy functions at 1000 °C to explain the structural findings of this study. The curves are deduced from proposed Gibbs free energy functions by Yashima et al. [89, 146]. The transformation c → t" → t" as observed by Yashima et al. [146] is indicated.

visualized by dark-field TEM imaging (Fig. 3.6). The size and density of the precipitates is independent of the location in the material, i.e., grain-core or grain-boundary regions.

As outlined in § 2.1.2 two metastable tetragonal phases, i.e., t'-YDZ and t"-YDZ, are known to form without diffusional processes in the relevant region of the phase diagram (cf. Fig. 4.1a). The diffusionless transformation of c-YDZ into the metastable t'-phase during rapid cooling of partially stabilized YDZ from the pure-cubic phase field is well described [73, 102, 103, 134–136, 144, 148, 193]. Larger t'-YDZ grains were reported to form during the diffusionless transformation of YDZ with Y_2O_3 contents up to 7 mol%. This is in contrast to the observation of homogeneously precipitated TNPs in this study. Zhou et al. [141] and Yashima et al. [89, 147, 148] observed the formation of the metastable t"-YDZ at dopant contents of about 7 mol% Y_2O_3.

From the crystallographic point of view t'-YDZ cannot be distinguished any more from t"-YDZ at 8.5 mol% Y_2O_3. The lattice parameters of t'-YDZ, which linearly depend on the dopant content, approach the lattice parameters of t"-YDZ and c-YDZ for this specific dopant content (cf. Fig. 2.5). Hence, no differences of the lattice parameters of the tetragonal phase and c-YDZ were resolved within the limits of SAED for 8YDZ-AS and the 10YDZ specimens [15].

Temporal fluctuations of the observed TNPs within the cubic matrix of the grains were shown by recording subsequent dark-field TEM images of the same sample region (cf. Fig. 3.8). This observation strengthen the assumption of a diffusionless formation of these regions. In addition, this instability indicate very small differences in Gibbs free energies between c-YDZ and the observed tetragonal-type phase at finite temperature irrespective of the cause. A small difference of the Gibbs free energy for c-YDZ and t"-YDZ (cf. Fig. 4.1b) was indeed proposed by Yashima et al. [89] for t"-YDZ.

Any change of the volume fraction of these regions, which would indicate an influence of the electron beam on crystal structure, could not be detected.

Hence, it is concluded that the homogeneously precipitated regions in 8YDZ-AS (cf. Fig. 3.6) are t"-YDZ. Although, the two metastable phase can not be distinguished at 8.5 mol% YDZ, it would be inappropriate to denote the observed tetragonal phase as t'-YDZ.

Comparison to 10YDZ

The microstructure of as-sintered and heat-treated 10YDZ (10YDZ-AS, 10YDZ-H) is very similar to the microstructure of 8YDZ-AS [15]. In both specimens TNPs were found, which are homogeneously distributed in the material. These regions are attributed to the formation of t"-YDZ like in 8YDZ-AS. This necessitates the expansion of the metastable c+t" two-phase field towards higher Y_2O_3 contents, which was proposed to be limited at lower dopant contents by Yashima et al. [148]. This expansion is indicated by the elongated t"-YDZ line (dashed line at lower temperatures) in Fig. 4.1a, which shows the metastable existence of t"-YDZ in YDZ specimens at room temperature.

As outlined in § 3.2.3, the volume fraction of these coherently embedded regions could not be derived from SAED patterns or dark-field TEM images. Nevertheless, both techniques indicate the significant decrease of the volume fraction of t"-YDZ precipitates with increasing Y_2O_3 content [15]. In contrast, the oxygen-vacancy concentration in YDZ increases with increasing dopant content. From these finding it is concluded that the TNPs are not necessarily associated with the local presence of an oxygen vacancy.

From the homogeneous precipitation as well as the temporal instability of the precipitates it is assumed that the material is chemically homogeneous as the 8YDZ-AS specimen. Hence, detailed analytical TEM analyses were not performed on the 10YDZ thick-film electrolyte specimens.

Up to now, it is not clarified completely, if the observed homogeneously distributed t"-YDZ precipitates exist in the 8YDZ and 10YDZ specimens during the applied heat treatments. Alternatively, they may form during cooling to room temperature. Yashima et al. [146] found the t"-YDZ formation in 7 mol% YDZ at very high temperature, i.e., 1400 °C. This is in accordance with the observations in the YDZ thin-film specimens as will be discussed later. At lower temperatures, the authors described the isothermal partial transformation of t"-YDZ into t'-YDZ and thus the coexistence of both phases in YDZ with dopant contents ≤7 mol%. This transformation was not found in the specimens investigated in this study.

4.1.2 Decomposition of 8YDZ

Coarsening of the TNPs in 8YDZ was observed at 950 °C (8YDZ-H) and 1000 °C (8YDZ-L1500, 8YDZ-L3000, and 8YDZ-L5800) by dark-field TEM imaging. The evolution is characterized by the formation of nearly spherical regions of about 10 nm size (cf. Fig. 3.7), which mainly consist of agglomerated TNPs. In these coarsened regions, the TNPs were observed to exist much more statically than in between these regions (cf. Fig. 3.9). This may indicate more pronounced differences of the Gibbs free energy.

The microstructural coarsening is accompanied by the distinct decomposition of

8YDZ. This was clearly demonstrated by combined EDXS/EELS experiments on the nanoscale (§ 3.2.4) as summarized in Fig. 3.27. It was found that the spherical coarsened regions are strongly depleted of Y ions. With respect to the measured Y concentration, it has to be taken into account that the coarsened tetragonal regions with sizes of about 10 nm are embedded in the TEM sample. Thus, the determined depletion of Y in these regions might be underestimated. The content of the Y-enriched surrounding YDZ material was only increased by a few 0.1 mol%. This is due to the small volume fraction of Y-depleted YDZ of about 10 vol%. This slightly increased dopant content in the Y-enriched YDZ is insufficient to completely stabilize the cubic high-temperature phase at room temperature since.

Heterogeneous or homogeneous nucleation and growth of t-YDZ precipitates with the equilibrium Y_2O_3 content (solvus) of a few at% (cf. Fig. 4.1a), which is well known for partially stabilized YDZ with lower Y_2O_3 contents [102, 113, 141, 148, 193], does not explain the microstructural evolution in the 8YDZ thick-film specimens. The size of the spherical coarsened regions even after 5800 h (1000 °C) is very similar to the the size of these regions after shorter heat treatments, e.g. in 8YDZ-H (cf. Fig. 3.7a and b). This finding contradicts the model of the nucleation in combination with the growth of t-YDZ. As an alternative, the observed decomposition of 8YDZ may be explained by spinodal decomposition.

Spinodal decomposition was previously found for YDZ with dopant contents up to 6 mol% as outlined in § 2.1.2. However, analytical data especially at higher Y_2O_3 contents is missing. Katamura and co-workers [119] and Shibata et al. [120] described the so-called "tweed"-contrast in bright-field TEM imaging due to the lamellar microstructure in 6 mol% YDZ specimens. A similar contrast (more speckle-like), attributed to strain variations in the material, contributes to bright-field TEM images of 8YDZ-H as shown in Fig. 3.10a [191]. The lamellar microstructure in 6 mol% YDZ results from the spinodal decomposition during heat treatment at 1200 °C for 126 h [120]. The authors state that the final microstructure consists of alternating c-YDZ and t-YDZ. They observed an increase of the local variations of the Y concentration in the lamellae with time by EDXS while the wavelength of the lamellae remained constant at about 15 nm [119]. Such increase of the compositional variations is typical for spinodal decomposition according to Cahn's theory [194]. Finally, the Y_2O_3 content within the lamellae was observed to alternate from about 2–4 mol% to about 8 mol% at the applied temperature of 1200 °C. From the observed solvus for Y in the enriched regions, i.e., about 8 mol%, Katamura and co-workers predicted spinodal decomposition to occur up to about 8 mol% Y_2O_3 at 1200 °C [119]. This solvus (marked in Fig. 4.1a) coincides well with the position of the revised phase boundary at this specific temperature. A value for the solvus of Y in c-YDZ at 1500 °C, i.e., 7.5 mol%, was proposed by Gibson & Irvine [144]. This value is also marked in Fig. 4.1a.

In the present study, the 8YDZ specimens were heat-treated at lower temperatures of about 1000 °C. This low temperature causes comparably low diffusivities for the Y and Zr ions and thus a longer time for 8YDZ to decompose. The average distance x, which the cations diffuse in time t can be estimated by an approximate solution for the second Fick's law [195]:

$$x = \sqrt{D \cdot t} = \sqrt{D_0 \cdot t \cdot e^{\left(-\frac{Q}{RT}\right)}} \qquad (4.1)$$

D_0 is a pre-exponential factor of the temperature-dependent diffusion coefficient, i.e., $3.1 \times 10^{-6}\,\mathrm{m^2 s^{-1}}$ for Zr ions [106, 109]. Q represents the activation energy for the transport of cations in the material. Using an activation energy of 4.5 eV [106–111] for Zr-ion transport in YDZ, the ratio of the diffusion coefficients and thus the ratio of the diffusion distances can be determined for two different temperatures. For 1200 °C, as in the study of Shibata et al. [120], and 1000 °C (this study), the ratio of the diffusion coefficients arises to 260. Hence, the ratio of the diffusion distances is about 16. This estimation clearly shows that the decomposition of the material is much slower at lower temperatures of around 1000 °C.

According to the lever rule, a significantly smaller volume of Y-depleted regions is expected for 8YDZ than for 6YDZ. 8YDZ is situated in the vicinity to the spinodal at higher Y concentrations, while 6YDZ is situated in the centre of the two-phase field. The estimated volume fraction of the coarsened regions in 8YDZ-L5800 arises to about 10 vol% (§ 3.2.3). This imbalance of the volume fractions of the final Y-depleted and the enriched regions explains the spherical shape of the separated Y-depleted regions found in 8YDZ-H and 8YDZ-L5800 in comparison to the lamellar structure in 6YDZ [119]. Nevertheless, the lateral size of the Y-depleted regions in 8YDZ-H and 8YDZ-L is on similar length scale as the width of the lamellae in 6YDZ observed by Katamura et al. [119]. This finding in addition strengthens the conclusion about the spinodal nature of the observed decomposition. Cahn's model [194] on spinodal decomposition predicts constant lateral dimensions of the compositional variations independent of time and initial composition of the decomposing material.

The variations of the local Y concentration in 8YDZ-H range from about 10 at% Y (5 mol% Y_2O_3) in the Y-depleted regions to about 18 at% (9 mol% Y_2O_3) in the enriched ones (Fig. 3.27). Assuming that the Y concentration in the depleted regions is underestimated due to signal averaging through the TEM sample, both values tend towards solvuses, predicted by the observations of Katamura et al. [119] and the phase diagram [89]. Since the slope of the t — c+t phase boundary is very large below 1500 °C (cf. Fig. 4.1a) a similar solvus for Y in t-YDZ has to be expected for 1200 °C [119] and for 1000 °C (this study). Hence, the solvus of Y in c-YDZ, i.e.,

$C_{\text{c-YDZ}}$, and thus the position of the c+t — c phase boundary at 1000 °C can be estimated using the lever rule:

$$\frac{m_{\text{t-YDZ}}}{m_{\text{c-YDZ}}} = \frac{C_{\text{c-YDZ}} - 8.52\,\text{mol}\%}{8.52\,\text{mol}\% - C_{\text{t-YDZ}}}$$

Here, $C_{\text{t-YDZ}}$ is the solvus of Y in t-YDZ, i.e., 2–3 mol% [73, 89, 97, 121]. Since the densities of t-YDZ and c-YDZ are very similar (cf. Table 2.1), the ratio of the volume fractions of both the Y-depleted and enriched regions, i.e., $\frac{0.1}{0.9}$ (cf. §3.2.3), instead of the ratio of the resulting mass fractions $\frac{m_{\text{t-YDZ}}}{m_{\text{c-YDZ}}}$ was used. Hence, the solvus for Y in c-YDZ at 1000 °C arises to about 9.2 mol% Y_2O_3. This is in excellent agreement with the predictions of Hanic et al. [165] who proposed the stabilization of the cubic phase in YDZ with dopant contents higher than 9.2 mol% (1000 °C). The determined solvus of 9.2 mol% at 1000 °C was used to construct the revised boundary line of the c+t two-phase field in Fig. 4.1a.

Any changes of the tetragonal lattice parameters that have to be expected from the formation of the t-phase could not be resolved within the detection limits of SAED. It is assumed that the formation of the equilibrium t-YDZ, which may has a lower Gibbs free energy (cf. Fig. 4.1b), is prevented in the studied specimens by the counteracting strain energy that would occur during the transformation in the coarsened tetragonal regions in 8YDZ-L5800. Hence, the supercooling of YDZ at room temperature in the dopant range of 7.3–10 mol% is insufficient for the nucleation of t-YDZ. Furthermore, strain may also lead to the peculiar constitution of the depleted regions consisting of both a tetragonal-type and the cubic phase in contrast to the lamellar structure with separated c-YDZ and t-YDZ in 6YDZ [119]. In the dark-field images of 8YDZ-H and 8YDZ-L5800 (Fig. 3.7) it can be recognized that the structure factors of the {112} reflection used for the imaging process may depend on local chemistry. While the contrast within coarsened regions and within the surrounding material in 8YDZ-H is similar it strongly varies in 8YDZ-L5800 (cf. Fig. 3.7a and 3.7b), which exhibits the advanced stage of decomposition. In Fig. 3.7b, the TNPs in the coarsened regions are much brighter than the precipitates in the surrounding material. This can be explained by the increased shift of oxygen-ion sites in the tetragonal precipitates within the coarsened and thus Y-depleted regions.

It has to be noticed that martensitic transformation of any tetragonal phase into the monoclinic phase [82] could not be observed applying SAED [15]. This is in accordance with observations of Lanteri et al. [112] and Ciacchi et al. [30] who ascribed this behaviour to the large volume increase, resulting in strain-field energy, that would accompany the phase transformation.

The decomposition on the cationic sublattice has to be accompanied by the rearrangement of the oxygen ions in the material due to charge neutrality. Any local separation of quasi-charged $V_O^{\bullet\bullet}$ and Y'_{Zr} species leads to electrical potentials and thus to an additional energetic term contributing to the overall energy of the crystal. Distinct variations of the local oxygen concentration were found by EELS in the distributions of the O-K signal (Fig. 3.27). The oxygen net signals along the line scans in the 8YDZ-H (Fig. 3.27b) run inversely to the local Y concentration as expected from charge neutrality. Furthermore, a more careful analysis of the oxygen distributions leads to the impression that oxygen vacancies accumulate in the vicinity of the Y-depleted regions. The small arrows in Fig. 3.27b mark regions with low O-K net signal, i.e., high $V_O^{\bullet\bullet}$ content. A second reason for the local decrease of the oxygen signal might be preferential electron-beam induced release of oxygen ions during the analytical experiments. This may result from varying binding energies or the influence of strain fields surrounding the coarsened regions. Up to now, such an artefact cannot be completely excluded. However, a drastic influence of the electron beam on the local composition during the measurement can be excluded because the intensity of the HAADF STEM images is homogeneous along the line scans immediately after the line-scan experiments as shown in Fig. 3.12b.

In conclusion, the coarsened regions in 8YDZ-H and 8YDZ-L5800 cannot be described in terms of the equilibrium t-YDZ phase. Furthermore, the constitution of the coarsened regions of tetragonal-type and cubic phase may be caused by the confinement in the surrounding material. Nevertheless, the decomposition of the material towards the equilibrium solvuses was found by analytical TEM. The TNPs in the Y-enriched surrounding material are explained by the formation of t"-YDZ, as it was found in 8YDZ-AS and 10YDZ. These TNPs are already present in 8YDZ-AS and even in 10YDZ.

4.1.3 Correlation of Chemical Decomposition and Decrease of Ionic Conductivity

The decrease of conductivity of YDZ with a dopant content of about 8–9 mol% is well-known for many years as outlined in § 2.1.1. From impedance spectroscopy it is concluded that the degradation of 8YDZ at higher temperatures is mainly caused by the increase of the bulk resistivity [11, 15, 31, 196]. The microstructural and chemical decomposition of bulk 8YDZ has been discussed as one reason for this decrease (cf. § 2.1.1).

Vlasov & Perfiliev [196] discussed the formation of a phase with low conductivity in addition to the original phase. However, this *low-conducting phase* is not further specified in detail in the manuscript. Ciacchi et al. [30] stated that the reduction of ionic conductivity in their studied 8 mol% YDZ specimen is caused by the precipitation of t-YDZ during annealing at 1000 °C for 2000 h (similar to 8YDZ-H, 8YDZ-L3000, 8YDZ-L5800). This conclusion is based on several observations. They had observed the formation of 6 nm large precipitates of an tetragonal-type phase after heat treatment (2000 h at 1000 °C) by dark-field TEM imaging, which they denoted as t-YDZ precipitates. Their published dark-field micrographs of the annealed 8 mol% YDZ specimen look similar to images of 8YDZ-H and 8YDZ-L5800 at intermediate magnification. At lower magnification the nanoscaled precipitates in the coarsened regions in 8YDZ-H and 8YDZ-L5800 (Fig. 3.7) appear as continuous tetragonal regions with sizes in the range of 5–10 nm. However, Ciacchi and co-workers [30,31] found a slight increase of the lattice parameter ($< 0.3\,\text{‰}$ by XRD using a Guinier fine focussing camera, Cu Kα_1) of the cubic matrix after the heat treatment. This increase was interpreted as the enrichment of Y in the cubic matrix of the grains due to the phase separation and thus the formation of t-YDZ with lower Y_2O_3 content. On the other hand, the authors stated that they had not detected any splitting of reflections neither by XRD nor SAED. In SAED, such splitting of reflections due to the differences of the lattice parameters of c-YDZ and t-YDZ (cf. Table 2.1) should be detectable. No information about the identification of t-YDZ is given in the manuscript. The chemical inhomogeneities in the material due to the formation of t-YDZ had not been shown by analytical TEM. Further information about tetragonal-type phases in the as-sintered specimens is not given by the authors.

Several groups utilized Raman spectroscopy for the characterization of the crystal structure of YDZ [12,148,197]. Hattori and co-workers [12] used Raman spectroscopy to identify structural changes of YDZ specimens with Y_2O_3 contents in the range of 8–10 mol%. Additional Raman bands were only observed for 8–9 mol% YDZ after heat treatment at 1000 °C for 1000 h. These were attributed to the formation of a tetragonal-type (not specified in detail) phase in the material [197], which is discussed as reason for the decrease of ionic conductivity. The specific Raman shifts of the maxima due to the tetragonal-type phase coincide well with the observations in this study.

In accordance with the findings of Hattori et al. [12] no indication for tetragonal-type Raman bands was detected for 8YDZ-AS (Fig. 3.28a). This shows that Raman spectroscopy is indeed suitable to detect coarsened tetragonal regions in heat-treated YDZ. However, it is inappropriate to detect the homogeneously distributed TNPs in as-sintered 8YDZ and 10YDZ.

Yashima et al. [148] found tetragonal-type Raman bands in YDZ specimens, containing up to 8 mol% Y_2O_3 which were prepared by arc melting. However, according to Ref. [148] the intensity ratios of these tetragonal-type Raman bands continously change with increasing Y_2O_3 content. Hence, an obvious conclusion regarding the existence and amount of tetragonal-type YDZ in these specimens cannot be drawn.

Furthermore, the complex crystal structure of YDZ in the range of Y_2O_3 contents around 8 mol% seems to impede the reliable utilization of Raman spectroscopy for structural analyses.

In the following, the experimental results regarding the chemical and structural decomposition of 8YDZ on the nanoscale are discussed in the context of the strong decrease of ionic conductivity in the material. As outlined in §2.1.1 (Fig. 2.1) 8–9 mol% YDZ exhibits the highest ionic conductivity in the system Y_2O_3–ZrO_2. The decomposition of 8YDZ on the nanoscale was found to lead to separated Y-depleted regions (Fig. 3.7) with dopant contents below 10 at% Y (Fig. 3.27) on the cationic sublattice while the Y_2O_3 content in the enriched surrounding material is slightly higher than the defined. The volume fraction of the depleted YDZ was determined to about 10 vol%.

Assuming a simple model of parallel conduction paths for the Y-depleted and the enriched YDZ results in a lower bulk conductivity σ_{decomp} for 8YDZ-H than for as-sintered 8YDZ if one averages the conductivities of the Y-depleted and Y-enriched regions according to $\sigma_{decomp} = 0.1 \cdot \sigma_{Y\text{-depleted}} + 0.9 \cdot \sigma_{Y\text{-enriched}}$. Depending on the degree of chemical decomposition, which is expected to increase with time, the conductivity of the depleted regions has to decrease during operation.

As outlined in §2.1.1 a lower conductivity of the Y-depleted regions can be explained by the reduced oxygen-vacancy and thus charge-carrier concentration. Furthermore, the strain field, surrounding the Y-depleted coarsened regions in 8YDZ-H and 8YDZ-L5800, may in addition impede the movement of oxygen ions through the YDZ material. Hence, the mobility of charge carriers has to be expected to be strongly influenced since the lattice of the as-sintered 8YDZ, which can be expected to be optimal for oxygen-ion transport, is distorted. However, variations of the local oxygen content and thus of the local oxygen-vacancy concentration were found in 8YDZ-H by EELS (Fig. 3.27). On the first view, these variations are complementary to the local Y concentration as discussed in the experimental part. This is necessarily expected from local charge neutrality. On the second view, the localized decrease of the oxygen signal was found in the vicinity of Y-depleted regions in 8YDZ-H. This decrease may indicate the accumulation of oxygen vacancies in the strain field of such coarsened regions. It can be speculated that this accumulation leads to higher binding energies and thus to a reduced mobility of these trapped oxygen vacancies. Clustering of vacancies was found by Kondoh et al. [33] using X-ray absorption fine-structure analysis (EXAFS). Maybe, these clusters are formed in the vicinity of the coarsened tetragonal regions in 8YDZ-H explaining the slightly higher vacancy concentration close to the coarsened regions.

This finally leads to the conclusion that both the concentration of free oxygen vacancies as well as the mobility of these vacancies, respectively, are strongly influenced by the decomposition of 8YDZ. Any local charge separation would lead to local electrostatical potentials and thus intrinsic electrical fields in the material. Such fields on the scale of about 10 nm should be detectable by electron holography.

As shown in a precious publication [15, 191], a contoured diffuse background can be detected in the studied 8YDZ and 10YDZ specimens. This diffuse background may be explained by short-range ordering and/or clustering of oxygen vacancies in the material as outlined in § 2.1.1. Such clusters are expected to exhibit higher binding energies for oxygen vacancies and are thus expected to lead to lower conductivity. Since the increasing intensity of this background with increasing dopant content was found by SAED, clustering may in addition contribute to the decrease of ionic conductivity in 8YDZ during heat treatment as predicted by several authors [9,14,32–34]. This process cannot be excluded by the experimental results of this study. However, the influence on ionic conduction is expected to be much smaller than the abovementioned one by the decomposition of the material.

As shown by García-Martin et al. [198] it is difficult to visualize the short-range ordering in 8YDZ by HRTEM while it may be easier for highly doped zirconia. The authors only found an *average fluorite lattice* in homogenized 7.5 mol% YDZ. HRTEM images of 8YDZ-AS and 10YDZ are similar to those presented by García-Martin et al. [198]. From dark-field TEM images of 8YDZ-AS and 10YDZ separated nanoscaled regions with tetragonal symmetry were expected. Instead, tetragonal intensity contributions to the HRTEM images were found to be continuously present in the whole field of view of an image. This is explained by the temporal instability of the nanoscaled t"-YDZ as it was found by dark-field TEM imaging. Signal averaging during the exposure of HRTEM images leads to the experimental observation of an average fluorite type lattice with very weak additional intensity due to the t"-YDZ precipitates. This also holds for the YDZ thin films, where TNPs were found. An example HRTEM image, in which the average fluorite type lattice can be seen along $\langle 110 \rangle$, is presented in Fig. 3.40.

4.2 Nanocrystalline YDZ

High-quality nano- and microcrystalline 7.3 mol% YDZ thin films with thicknesses around 400 nm were investigated in this study. The prepared thin films were free of cracks with excellent adhesion on the sapphire substrates, independent of the final annealing temperature. This is in contrast to recent studies, in which comparable chemical deposition techniques were utilized by the authors to prepare nanocrystalline 8–8.5 mol% YDZ thin films [51, 58, 60, 62, 65] on various polycrystalline as well as single-crystalline substrates. These publications clearly demonstrate that the preparation of crack-free YDZ thin films exhibiting good adhesion is still challenging. No preferred crystal orientation within the YDZ thin films with respect to the sapphire substrates was found. This is in contrast to the observations of Miller & Lange [199] who utilized preferential (regarding the crystal orientation) growth of YDZ on sapphire substrates to prepare epitaxial films by spin-coating.

Most microstructural properties like film thickness, mean grain size, and porosity as well as crystal structure and chemical homogeneity of the specimens are relevant for the interpretation of the electrical characterization [16, 167]. Nevertheless, the characterization of the distribution of porosity for example even in chemically deposited 8–8.5 mol% YDZ thin films is often missing in literature [51,61,62,65,68,200]. As outlined in the fundamentals (§ 2.1.1), most authors [45–52,54–56,58–61,63,64,66,68,69, 201] limit the microstructural characterization of their studied 8–9.5 mol% YDZ thin-film specimens to easy-to-apply techniques, e.g. SEM and XRD. In contrast to TEM, SEM is an surface-sensitive technique restricted to the investigation of the surface-near microstructural properties of the prepared thin-film specimens. As shown for the nanocrystalline thin-films YDZ-650 and YDZ-850 that exhibit a layered morphology the distribution of pores may vary within specimens prepared by spin- or dip-coating. Such variations in nanocrystlline materials may be difficult to detect by SEM even in cross-section samples.

Comprehensive structural and microstructural characterization by TEM is relative sparse [67, 200]. This explains, why most authors [46, 49, 51, 52, 57, 59, 60, 63, 68, 201] detect only c-YDZ in their 8–9.5 mol% YDZ specimens by XRD. It is doubtful that the specimens in all these studies are pure cubic while other crystal structures are well-known for microcrystalline 8.5 and 10 mol% YDZ. However, the possible decomposition of YDZ during preparation of thin films makes a thorough characterization of the microstructure and chemistry on the nanoscale indispensable.

In the following discussion the dependency of grain growth in the thin films on the final annealing temperature is briefly raised. The main emphasis is on the crystal structure and distribution of phases in the thin films. The alternative to the sintering of green bodies is the used chemical deposition technique, for which the specimens are maximal heated up to the temperature of the final heat treatment. Hence, these specimens yields additional information on the phase diagram in the targeted dopant range.

4.2.1 Microstructure

The grain size variations in the range of 5 nm to nearly 1 μm in this work is considerably larger than in other studies on YDZ thin films [46, 47, 49–52, 59, 63, 64, 66, 67, 69, 199–205]. The studied YDZ thin-film specimens exhibit unimodal grain-size distributions indicating normal grain growth (Fig. 3.33). The resulting mean grain sizes ranging from 5 nm up to 1 μm can be controlled by the thermal treatment. Porosity of up to 15 vol% (Table 3.3) with pore sizes equal to the corresponding grain size is present in the samples annealed at temperatures ≤1000 °C (Fig. 3.34). The presence of pores after the final heat treatment can be understood by considering the preparation method based on a sol-gel precursor with a considerable volume fraction of organic material. During the RTA process the organic components of the precursor are pyrolysed resulting in pores in the samples YDZ-650, YDZ-850, and YDZ-1000. Upon heat treatment the YDZ layers densify, accompanied by grain growth, resulting in completely dense YDZ films for annealing temperatures exceeding 1000 °C. In comparison with published data on grain sizes of sol-gel derived YDZ thin films [49, 52, 54, 60, 200] the grain sizes of the nanocrystalline specimens (YDZ-650 – YDZ-1000) are slightly smaller. Only Gorman & Anderson [200] found similar grain sizes after comparable heat treatment of their 70 nm thick 8 mol% YDZ thin films.

Grain Growth

Grain growth usually follows a potential law. Consequently, Fig. 4.2 shows an Arrhenius plot of the logarithm of the average grain size and the relative density of the YDZ thin films as a function of the reciprocal annealing temperature. The region with porous thin films is shaded on the right side of the diagram. The slope of the graph of $\ln(d/[\text{nm}])$ represented by the straight line for the porous thin films changes with the annealing temperature. This slope is directly correlated with the growth rate. While grain growth seems to be limited in the porous thin films, it is strongly accelerated in the dense samples YDZ-1250 and YDZ-1350. Assuming negligible size of the YDZ grains in the YDZ thin films immediately after pyrolysis, grain growth can be described by the equation $d^n = D_0 t e^{(-\frac{Q}{RT})}$ [206]. In this equation d represents the grain diameter. Q is the activation energy for the predominant transport mechanism. t is the duration for the heat treatment at temperature T. The exponent n was found to vary significantly between 2 and 4 depending on the predominant transport mechanisms in ceramic materials. This exponent cannot be determined on the basis of the experiments in this study. This is due to the fact that only the temperature was varied using the same annealing time for all prepared YDZ thin-film specimens. Hence, the different transport processes for cations, e.g. surface diffusion in porous

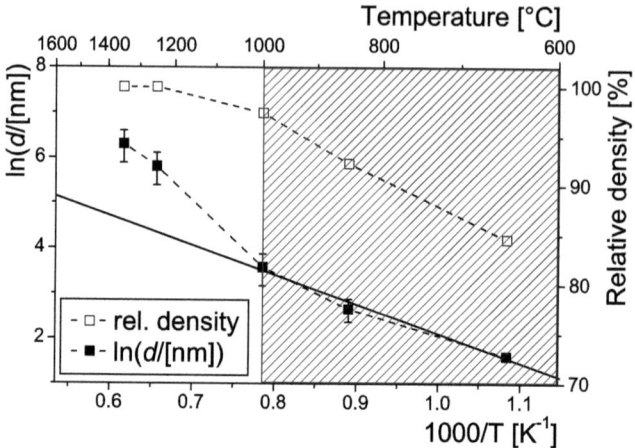

Figure 4.2: Grain growth in YDZ thin films. The straight and dashed lines are guides to the eyes.

thin films or diffusion in the bulk, during the densification of the thin films cannot be distinguished.

Plotting $\ln(d/[\mathrm{nm}])$ as a function of $1/T$ in Fig. 4.2 yields a straight line for the three porous specimens. This indicates that the same processes dominate the grain growth in these specimens. An acceleration of grain growth is observed for YDZ-1250 and YDZ-1350, which do not contain pores anymore after annealing for 24 h. These samples completely densify and coarsen during the heat treatment, which implies a possible change of both n and Q.

A time-dependent study with samples annealed at a specific temperature but for different times is necessary to clarify any change of the predominant transport mechanisms and thus the activation energy. Furthermore, information on the self-limitation of the grain growth could be derived as observed in nanocrystalline ceramics, e.g. in nanocrystalline ceria [207].

The limitation of grain growth during the densification of microcrystalline ceramic materials is well known and discussed frequently, e.g. by Yan [208]. The acceleration in grain growth concurrent with full densification for 3YSZ films was observed by Díaz-Parralejo et al. [202]. In specimen YDZ-1600, the disrupture of the film is driven by the interface energies in the system film-substrate-atmosphere as discussed by Sakurai et al. [203]. From a thermodynamic point of view, at lateral grain sizes larger than the film thickness, disruption of the film occurs because isolated grains become more stable than a polycrystalline film.

4.2.2 Chemistry

Grain boundaries in nanocrystalline YDZ electrolytes deserve particular attention because they might strongly affect the conductivity of SOFC electrolytes. Amorphous grain-boundary films due to segregating impurities like alumina or silica may have a negative impact on ionic conductivity in YDZ as discussed by Aoki et al. [160] and Kondoh et al. [33]. However, extensive EDXS analyses confirmed within the detection limits the absence of contaminants in the sol-gel layers investigated in this study. Applying HRTEM the existence of extensive amorphous phases at the grain boundaries is excluded. Previous studies confirmed that amorphous grain-boundary films may act as sinks for yttrium ions [155, 156], leading to the local segregation of yttrium ions in grain-boundary regions. In this case a local increase of the TNPs in the vicinity of grain boundaries must be expected from the results of the thick-film electrolytes, where the decreasing volume fraction of TNPs with increasing Y_2O_3 content was found. However, any local change of the density of the TNPs at grain boundaries was excluded by experimental observations as shown in Fig. 3.36c. This result is considered as a further confirmation for the absence of impurities at the grain boundaries. Significant chemical reaction of the YDZ with the sapphire substrates was not observed, although intensive analytical investigations were performed across the film/substrate interface and at the grain boundaries, which may act as potential diffusion path. The locally increased density of TNPs close to the grain boundaries as well as at the interface to the substrate (Fig. 3.37) may be ascribed to the in-diffusion of Al into the bulk and at the grain boundaries below the detection limitations of analytical TEM. Fabrichnaya & Aldinger [152] predicted a low solubility of Al in t-YDZ at high temperature, i.e., 1250 °C. However, the metastability of tetragonal-type YDZ has not been considered in this theoretical work about the ZrO_2-Y_2O_3-Al_2O_3 system. Hence, one can only speculate about the stabilization of t"-YDZ by a small amount of Al, which would lead to the observed increased density of t"-YDZ precipitates.

In contrast, Kosacki et al. [49, 52] ascribed the formation of m-YDZ in sol-gel derived 8 mol% YDZ thin films to the reaction with the sapphire substrates at temperatures above 1100 °C. This has not been confirmed in this study. The changes of the peak widths and heights of cubic-type reflections, found by Kosacki et at. [49, 52], may be correlated with the decomposition of their YDZ specimens heat-treated at the higher temperatures.

4.2.3 Crystal Structure

In the following the properties of the observed tetragonal-type phase in the sol-gel derived thin films are discussed and compared with the findings obtained from the

8YDZ thick-film electrolytes. The evaluation of the properties of the tetragonal-type YDZ is of particular interest in the regime of relatively low temperatures (1250 °C and below) because the specimens do not require high-temperature homogenization as a consequence of the fabrication method and solidification at low temperature. The slightly lower dopant content of the thin films in comparison to the 8YDZ and 10YDZ thick-film electrolytes yields additional information on the position of the c — c+t phase boundary.

SAED patterns of all thin-film specimens (Fig. 3.35) show, in addition to the reflections of c-YDZ, reflections, which can be clearly ascribed to YDZ with a tetragonal structure. This is in contrast to the observations of Gorman & Anderson [200] who studied similar nanocrystalline 8 mol% YDZ thin films prepared by spin coating using a polymeric precursor. However, the intensity of {112} reflections is extremely weak according to Debye SAED performed in this study (cf. Fig. 3.35b). Hence, it seems more likely that additional intensities due to tetragonal-type YDZ were overseen by Gorman & Anderson [200].

The lattice parameters of this non-cubic phase do not differ with respect to c-YDZ within the accuracy of SAED. A high density of nanoscaled YDZ inclusions with tetragonal structure and sizes around 1 nm were found to exist coherently embedded in the cubic matrix of the grains (Fig. 3.36 and 3.37). This microstructure is very similar to the microstructure found in the as-sintered 8YDZ and 10YDZ thick-film electrolytes (cf. Fig. 3.36 and 3.37 with Fig. 3.6).

4.2.4 Decomposition

While these precipitates tend to agglomerate in YDZ-1350, the distributions in all other studied thin films are homogeneous. The coarsening in YDZ-1350 is interpreted as indication for the incipient decomposition of the material as it was observed in 8YDZ-H and 8YDZ-L5800. The diffusion distances of cations at 1350 °C (24 h) by Eq. 4.1 correspond to a few nm. They are on the Å-scale for YDZ-650 – YDZ-1250 due to the lower annealing temperatures. This indicates that the time of 24 h is too short for the specimens YDZ-650 – YDZ-1250 to show significant inhomogeneities in contrast to YDZ-1350. Nevertheless, it is expected that these YDZ thin-film specimens are situated in the c+t two-phase field as indicated by the grey symbols in Fig. 4.1a.

A local reduction of the Y concentration was found by spot EDXS analyses in YDZ-1350. This in addition strengthens the assumption of the decomposition of YDZ-1350. Compositional inhomogeneities were not found in the other YDZ thin-film specimens.

Hence, the observed microstructure of YDZ-650, YDZ-850, YDZ-1000, YDZ-1250, and YDZ-1600 is interpreted in terms of the precipitation of t"-YDZ as it was found for the as-sintered 8YDZ and 10YDZ thick-film electrolytes. The observation of t"-YDZ precipitates in YDZ-1600 must be related to the formation of this phase during the cool-down process.

The TNPs are not detectable by XRD [189] due to the strong broadening of the tetragonal-type reflections. This finding is of considerable practical importance. For the interpretation of conduction phenomena in YDZ in the targeted range of Y_2O_3 contents, the amount and distribution of the tetragonal-type phase may has to be considered. Inhomogeneities in the distribution of the nanoscaled precipitates may be an indication for the decomposition or compositional inhomogeneity of the material as found in the 8YDZ thick-film electrolytes after SOFC operation.

4.3 Formation and Stability of Phases near c+t — c Phase Boundary

In the dopant range of 7.3–10 mol% Y_2O_3, homogeneously distributed, nanoscaled t"-YDZ precipitates were found in all studied specimens irrespective of the preparation route. The homogeneous distributions of these t"-YDZ precipitates in most 7.3 mol% YDZ thin-film specimens (except YDZ-1350) are similar to the distributions in 8YDZ-AS and 10YDZ. This clearly shows the consistency of the experimental results regarding the formation of t"-YDZ in the targeted dopant range. Furthermore, the volume fraction of t"-YDZ next to the cubic phase was found to decrease with increasing Y_2O_3 content. 16YDZ was found to be pure cubic. No indication for the formation secondary phases, e.g. of t"-YDZ, was found in this specimen. This demonstrates that the TNPs in the specimens with dopant contents in the range of 7.3–10 mol% Y_2O_3 are with high probability not an artefact induced by the electron beam.

The decomposition of YDZ was found to occur up to higher Y_2O_3 contents, as assumed in literature. This is indicated in the revised phase diagram (Fig. 4.1a). While 8YDZ-H needs several thousand hours at about 1000 °C, the YDZ thin-film specimen YDZ-1350 shows significant indications for the decomposition already after 24 h [192]. This is consistent with thermally activated cation diffusion in the YDZ material, which is strongly increased at the higher temperature. Since the thin films were only heat treated for a very short time (24 h), the decomposition of the thin-film specimens annealed at lower temperatures than 1350 °C is expected in much longer time.

The nucleation and growth of t-YDZ, as observed at lower Y_2O_3 contents, is suppressed in all specimens. This is attributed to the strain-field energy and insufficient supercooling of the specimens impeding the formation of stable nuclei of t-YDZ. Furthermore, t"-YDZ, which was observed to form easily, exists in the material, which in addition minimizes the overall energy of the material.

Indeed, Gibbs free energy functions, as proposed by Yashima et al. [89, 146], can explain the experiment findings of the this study. Gibbs free energy functions for $T = 1000\,°C$ are presented in Fig. 4.1b in addition to the revised phase diagram. The curves are deduced from the proposed Gibbs free energy functions presented by Yashima et al. [89, 146]:

- Formation of t"-YDZ in all studied specimens with dopant contents in the range of 7.3–10 mol% Y_2O_3: Narrow Gibbs free energy functions of c- and t"-YDZ (cf. Fig. 4.1b) in the c+t" two-phase field [148] explain the existence of both phases at finite temperatures. Moreover, the small size of the t"-YDZ precipitates can be explained considering the limited volume fraction of t"-YDZ in the c+t" two-phase field. Due to the small energetic differences of c- and t"-YDZ, of t"-YDZ is facilitated to nucleate homogeneously in all specimens (7.3–10 mol% Y_2O_3). The easy formation of t"-YDZ in combination with the limited volume fraction leads to the small size of the t"-YDZ precipitates.

- Temporal instability of t"-YDZ: The small energetic differences between c-YDZ and t"-YDZ may in addition explain the temporal instability of the t"-YDZ precipitates. Since the thermal energy $k_B T$ at room temperature is very low, i.e., 25 meV, the transformation during TEM investigation may be induced by the electron beam.

- Increasing volume fraction of t"-YDZ with decreasing Y_2O_3 content: This is explained by the differences of the slopes of the Gibbs free energies for t"-YDZ and c-YDZ with decreasing Y_2O_3 content. This leads to increasing energetic difference between both phases with decreasing Y_2O_3 content resulting in an increasing volume fraction of t"-YDZ.

- Transition of t"-YDZ into t'-YDZ: This transition was found by Yashima et al. [89, 146] for YDZ with lower Y_2O_3 contents up to 7 mol% Y_2O_3 as indicated in Fig. 4.1b. In the present study the range of Y_2O_3 contents (7.3–10 mol% Y_2O_3) is too high to facilitate this transition. This means the overall energy of the material consisting of c-YDZ and t"-YDZ falls below the Gibbs free energy of t'-YDZ as proposed by Yashima et al. [89,146]. Hence, both Gibbs free energy functions for t"-YDZ and c-YDZ are below the Gibbs free energy function of t'-YDZ.

- Missing nucleation and growth of equilibrium t- and c-YDZ in the targeted dopant range: The flat slope of the Gibbs free energy function of the stable t-YDZ in combination with the higher slopes of the Gibbs free energy functions of c-YDZ as well as t"-YDZ around 8.5 mol% Y_2O_3 impede the nucleation and growth of the stable equilibrium phases. Instead, a mixture of c- and t"-YDZ minimizes the crystal energy in this range of Y_2O_3 contents, i.e., 7.3–10 mol%. Furthermore, the supercooling of the material in the studied dopant range is expected to be insufficient to facilitate the formation of stable nuclei of the equilibrium phase. This is attributed to additional energetic contributions like strain-field energy, which in addition impede the nucleation.

The observation of a homogeneous distribution of TNPs in YDZ-1600 as well as in the as-sintered thick-film specimens 8YDZ-AS, 8YDZ-ASH, and 10YDZ demonstrates that the complete stabilization of the cubic phase can be excluded with high probability for YDZ with Y_2O_3 contents up to 10 mol%.

Summary

Y_2O_3-doped ZrO_2 (YDZ) has been the standard solid electrolyte for solid oxide fuel cells for many years. Nevertheless, important questions regarding the ionic conductivity are still controversially discussed in literature. This concerns in particular YDZ with dopant contents between 7 and 10 mol% Y_2O_3. The focus of this work was the microstructural, structural, and chemical characterization of YDZ specimens by various TEM techniques to contribute to the clarification of these questions.

Thick-Film Electrolytes

To understand the strong decrease of ionic conductivity of 8YDZ at high operation temperatures (this study: nearly 40 % after 2500 h at 950 °C), polycrystalline 8YDZ electrolyte foils (thickness 200 µm) were analysed before and after operation. In contrast, negligible degradation of the initial conductivity was observed for 10 mol% YDZ (10YDZ) at the same temperature after 1000 h. Impedance spectroscopy demonstrates that changes at grain boundaries (e.g. formation of a less-conducting glassy phase) are not responsible for the degradation of ionic conductivity of the high-purity 8YDZ specimens in this study. Therefore, changes of the bulk properties must reason the decrease of conductivity. A proposed explanation is the decomposition of 8YDZ in the c+t two-phase field of the Y_2O_3–ZrO_2 phase diagram. This concerns in particular YDZ with dopant contents between 7 and 10 mol% Y_2O_3.

A metastable tetragonal phase (t"-YDZ) was observed in the as-sintered 8YDZ and 10YDZ specimens, coherently embedded in the cubic matrix. The distribution of t"-YDZ was visualized by dark-field TEM imaging using {112} reflections of the tetragonal YDZ phase. t"-YDZ homogeneously precipitates in nanoscaled regions (size $\sim$1 nm) with a density that does not depend on grain-core and grain-boundary regions. Temporal fluctuations of the t"-YDZ precipitates were detected in-situ during TEM investigation indicating a low energetic barrier between both phases.

In 10YDZ, which does not show degradation of ionic conductivity, no changes of the distribution of t"-YDZ were found after the heat treatment. In contrast, the local coarsening of the tetragonal nanoscaled regions was observed for 8YDZ after heat treatment (2500 h at 950 °C). The size of the coarsened regions in 8YDZ is about 10 nm.

Applying quantitative EDXS and EELS line-scan measurements, the chemical decomposition of 8YDZ on the nanoscale was clearly shown after degradation. Analytical TEM studies of as-sintered 8YDZ reveal a homogeneous composition within the detection limits of the techniques. The coarsened regions, occupying about 10 vol% in 8YDZ after decomposition, are significantly depleted in dopant cations while the surrounding material is slightly enriched in Y. The Y_2O_3 content alternates between 4 mol% in the Y-depleted and 10 mol% in the Y-enriched 8YDZ. Hence, it tends towards the solvuses proposed for the equilibrium t- and c-YDZ phases according to the phase diagram. The local oxygen content was observed to run inversally in comparison to the local Y content. This is expected from charge neutrality. Furthermore, the distinct decrease of the local oxygen signal was detected in the vicinity to the coarsened regions by quantitative EELS, indicating the accumulation of oxygen vacancies close to the coarsened regions in heat-treated 8YDZ.

The variations of the local Y and O concentrations can be correlated with the reduction of ionic conductivity. The Y-depleted regions are characterized by a reduced content of oxygen vacancies. The accumulation of the oxygen vacancies in the vicinity of the coarsened regions may indicate the reduction of the mobility of the vacancies. Hence, the mobility as well as the concentration of free oxygen vacancies are expected to be affected negatively by the decomposition of the material.

However, the formation of the equilibrium t-YDZ in the Y-depleted regions could not be confirmed by electron diffraction. The relaxation of coarsened Y-depleted regions into equilibrium t-YDZ may be prevented by the confinement in the Y-enriched cubic matrix, which would lead to a large strain-energy contribution.

YDZ Thin Films

Nano- and microcrystalline 7.3 mol% YDZ (7YDZ) thin films (thicknesses ~400 nm) were studied to gain information about the electrical properties of YDZ on the nanoscale. In particular, the contribution of grain boundaries to the overall conductivity is of interest. Therefore, thin films of high purity were prepared by dip-coating sapphire substrates using a sol-gel precursor. Grain-size variation within the series of specimens was achieved by an additional heat treatment after pyrolysis in the temperature range of 650-Ű1600 °C for 24 h, resulting in homogeneous YDZ thin films with mean grain sizes in the range of 5 nm to about 0.5 µm. Despite the low-temperature preparation method, well-crystallized YDZ thin films were obtained already at 650 °C.

The nanocrystalline specimens (T ≤ 1000 °C) exhibit remaining porosity. In these specimens grain growth appears to be limited while it is accelerated in the thin films annealed at higher temperatures. EDXS did not yield any indication for impurities in the grains nor segregation at the grain boundaries, thus confirming chemical homogeneity and low impurity content of the investigated YDZ thin films.

As in the thick-film electrolytes with slightly higher dopant contents, t"-YDZ was found to precipitate homogeneously in all studied thin films. The coarsening of the nanoscaled regions (as in 8YDZ after operation) was also observed incipiently in the 7YDZ thin film, heat treated at 1350 °C (24 h). The microstructural coarsening indicates compositional inhomogeneities in the material due to decomposition, which were confirmed by local EDXS measurements.

The variation of the average grain size by two orders of magnitude results in a significant variation of bulk-to-grain boundary volume, thus enabling thorough studies of the ionic conductivity with respect to the influence of grain boundaries. The series of films showed a decrease of electrical conduction upon grain-size reduction. This is attributed to poor charge transport at grain boundaries, which are characterized by a specific conductivity ranging approximately two decades below the specific bulk conductivity. Noteworthy, the specific grain-boundary conductivity is constant in the thin films with mean grain sizes $d \geq 232$ nm.

Y_2O_3–ZrO_2 Phase Diagram

The investigation of YDZ with doping contents between 7.3 and 10 mol% Y_2O_3 yields new information regarding the Y_2O_3–ZrO_2 phase diagram. In contradiction to the assumption of the stabilization of 8YDZ in the cubic phase, a metastable tetragonal phase (t"-YDZ) was found to form in all specimens up to 10 mol% Y_2O_3 irrespective of the preparation route. Furthermore, the microstructural and chemical decomposition of 8YDZ and 7YDZ thin films was observed by analytical TEM, while 10YDZ was stable at 950 °C.

Metastable t"-YDZ Nanoscaled t"-YDZ precipitates were observed in all as-prepared YDZ specimens with of 7.3–10 mol% Y_2O_3. Despite different preparation routes, i.e., sintering of green bodies (thick-film electrolytes) vs. dip-coating using a sol-gel precursor (thin films), the crystal structure as well as the distributions of t"-YDZ in the materials are similar. This indicates the consistency of the observations regarding the formation of t"-YDZ. In contrast, doping with 15.6 mol% Y_2O_3 leads to single-phase cubic YDZ. The volume fraction of t"-YDZ was found to significantly decrease with increasing dopant content up to 10 mol% Y_2O_3.

The formation of t"-YDZ is described in literature for dopant contents up to 7–8 mol%

Y_2O_3. However, the distribution of t"-YDZ in YDZ and the resulting microstructure was not visualized by TEM yet. In contrast to SAED, the t"-YDZ precipitates cannot be detected in the studied specimens by XRD despite the significant volume fraction of the phase. This is attributed to the small size of the nanoscaled regions and the small structure factors of the additional reflections of the t"-YDZ phase.

The findings of this study clearly necessitate the expansion of the coexistence field of c+t"-YDZ towards higher dopant contents up to at least 10 mol% Y_2O_3. Furthermore, qualitative Gibbs free energy functions for the applied annealing temperature are deduced from literature data, which explain the decrease of the volume of this phase with increasing dopant content. The observation of t"-YDZ up do dopant contents of 10 mol% Y_2O_3, which cannot be detected by XRD and Raman spectroscopy, excludes with high probability the stabilization of c-YDZ at room temperature up to 10 mol% Y_2O_3.

c+t Two-Phase Field A revised position of the boundary between the c+t and c phase field was derived. The decomposition of YDZ was shown for 7YDZ thin films (for temperatures $\leq 1350\,°C$) as well as for 8YDZ (950 °C). No changes were found for 10YDZ at 950 °C. While 8YDZ necessitates long time (few 1000 h) to decompose, the 7YDZ thin-film specimen annealed at 1350 °C showed incipient decomposition even after 24 h. This is explained by the temperature dependency of the cation diffusivity in YDZ. Since no microstructural changes were found for the materials at higher temperatures, a revised position of the c+t — c phase boundary at higher Y_2O_3 contents could be defined in the investigated dopant range.

The decomposition in the c+t two-phase field is discussed in terms of spinodal decomposition, since the lateral size of the Y-depleted regions did not change significantly if the annealing duration was extended from 2500 h to 5800 h (1000 °C).

Indeed, lowering the operation temperature to $T \leq 800\,°C$ slows down the process of decomposition to a minimum without the need to increase the Y_2O_3 concentration to obtain completely stabilized YDZ.

Bibliography

[1] P. HOLTAPPELS and U. STIMMING. *Handbook of Fuel Cells: Fundamentals, Technology and Applications*, Vol. 1, Ch. Solid Oxide Fuel Cells (SOFC), (pp. 335–354) (John Wiley & Sons Ltd., Chichester, England, 2003).

[2] S. SINGHAL and K. KENDALL. *High-temperature Solid Oxide Fuel Cells: Fundamentals, Design and Applications*. First ed. (Elsevier Ltd., New York, 2003).

[3] E. RYSHKEWITCH. *Oxide ceramics. Physical Chemistry and Technology* (Academic Press, New York and London, 1960).

[4] K. K. SRIVASTAVA, R. N. PATIL, C. B. CHOUDHARY, K. V. G. K. GOKHALE and E. C. SUBBARAO. *Revised Phase Diagram of the System ZrO_2–$YO_{1.5}$*. Brit. Ceram. Trans. J., Vol. 73(1), pp. 85–91 (1974).

[5] A. H. HEUER, R. CHAIM and V. LANTERI. *Review: Phase Transformations and Microstructural Characterization of Alloys in the System Y_2O_3-ZrO_2*. S. SOMIYA, N. YAMAMOTO and H. YANAGIDA (Eds.), *Advances in Ceramics: Science and Technology of Zirconia III*, Vol. 24, (pp. 3–20) (American Ceramics Society, Westerville, Ohio, 1988).

[6] M. RÜHLE. *Microscopy of Structural Ceramics*. Adv. Mater., Vol. 9(3), pp. 195–217 (1997).

[7] R. C. GARVIE, R. H. HANNINK and R. T. PASCOE. *Ceramic steel?* Nature, Vol. 258, pp. 703–704 (1975).

[8] W. NERNST. *Über die Elektrolytische Leitung fester Körper bei sehr hohen Temperaturen*. Z. Elektrochem., Vol. 6(2), pp. 41–43 (1899).

[9] J. KONDOH, T. KAWASHIMA, S. KIKUCHI, Y. TOMII and Y. ITO. *Effect of Aging on Yttria-Stabilized Zirconia: I. A Study of Its Electrochemical Properties*. J. Electrochem. Soc., Vol. 145(5), pp. 1527–1536 (1998).

[10] C. C. APPEL, N. BONANOS, A. HORSEWELL and S. LINDEROTH. *Ageing behaviour of zirconia stabilised by yttria and manganese oxide*. J. Mater. Sci., Vol. 36(18), pp. 4493–4501 (2001).

[11] N. BALAKRISHNAN, T. TAKEUCHI, K. NOMURA, H. KAGEYAMA and Y. TAKEDA. *Aging Effect of 8 mol % YSZ Ceramics with Different Microstructures.* J. Electrochem. Soc., Vol. 151(8), pp. A1286–A1291 (2004).

[12] M. HATTORI, Y. TAKEDA, Y. SAKAKI, A. NAKANISHI, S. OHARA, K. MUKAI, J.-H. LEE and T. FUKUI. *Effect of aging on conductivity of yttria stabilized zirconia.* J. Power Sources, Vol. 126(1), pp. 23–27 (2004).

[13] M. HATTORI, Y. TAKEDA, J.-H. LEE, S. OHARA, K. MUKAI, T. FUKUI, S. TAKAHASHI, Y. SAKAKI and A. NAKANISHI. *Effect of annealing on the electrical conductivity of the Y_2O_3–ZrO_2 system.* J. Power Sources, Vol. 131(1–2), pp. 247–250 (2004).

[14] C. HAERING, A. ROOSEN and H. SCHICHL. *Degradation of the electrical conductivity in stabilised zirconia systems: Part I: yttria-stabilised zirconia.* Solid State Ionics, Vol. 176(3–4), pp. 253–259 (2005).

[15] B. BUTZ, P. KRUSE, H. STÖRMER, D. GERTHSEN, A. MÜLLER, A. WEBER and E. IVERS-TIFFÉE. *Correlation between microstructure and degradation in conductivity for cubic Y_2O_3-doped ZrO_2.* Solid State Ionics, Vol. 177(37–38), pp. 3275–3284 (2006).

[16] C. PETERS. *Grain-size effects in nanoscaled electrolyte and cathode thin films for solid oxide fuel cells (SOFC).* PhD thesis, University of Karlsruhe (TH) (2009).

[17] F. A. KRÖGER and H. J. VINK. *Relations between the concentrations of imperfections in solids.* J. Phys. Chem. Solids, Vol. 5(3), pp. 208–223 (1958).

[18] F. HUND. *Anomale Mischkristalle im System ZrO_2–Y_2O_3. Kristallbau der Nernst-Stifte.* Zeitschrift für Elektrochemie und Angewandte Physikalische Chemie, Vol. 55(5), pp. 363–366 (1951).

[19] D. W. STRICKLER and W. G. CARLSON. *Ionic Conductivity of Cubic Solid Solutions in the System CaO–Y_2O_3–ZrO_2.* J. Am. Ceram. Soc., Vol. 47(3), pp. 122–127 (1964).

[20] J. M. DIXON, L. D. LAGRANGE, U. MERTEN, C. F. MILLER and J. T. PORTER II. *Electrical Resistivity of Stabilized Zirconia at Elevated Temperatures.* J. Electrochem. Soc., Vol. 110(4), pp. 276–280 (1963).

[21] R. E. W. CASSELTON. *Low Field DC Conduction in Yttria-Stabilized Zirconia.* Phys. Status Solidi A, Vol. 2(3), pp. 571–585 (1970).

[22] I. R. GIBSON, G. P. DRANSFIELD and J. T. S. IRVINE. *Influence of Yttria Concentration upon Electrical Properties and Susceptibility to Ageing of Yttria-stabilised Zirconias*. J. Eur. Ceram. Soc., Vol. 18(6), pp. 661–667 (1998).

[23] J. A. KILNER and C. D. WATERS. *The Effects of Dopant Cation-Oxygen Vacancy Complexes on the Anion Transport Properties of Non-Stoichiometric Fluorite Oxides*. Solid State Ionics, Vol. 6(3), pp. 253–259 (1982).

[24] J. F. BAUMARD and P. ABELARD. *Defect Structure and Transport Properties of ZrO_2-Based Solid Electrolytes*. M. RÜHLE, N. CLAUSSEN and A. H. HEUER (Eds.), Advances in Ceramics: Science and Technology of Zirconia II, Vol. 12, (pp. 555–571) (American Ceramics Society, Columbus, Ohio, 1984).

[25] A. NAKAMURA and J. B. W. JR. *Defect Structure, Ionic Conductivity, and Diffusion in Yttria Stabilized Zirconia and Related Oxide Electrolytes with Fluorite Structure*. J. Electrochem. Soc., Vol. 133(8), pp. 1542–1548 (1986).

[26] O. YAMAMOTO, Y. ARACHI, H. SAKAI, Y. TAKEDA, N. IMANISHI, Y. MIZUTANI, M. KAWAI and Y. NAKAMURA. *Zirconia Based Oxide Ion Conductors for Solid Oxide Fuel Cells*. Ionics, Vol. 4(5-6), pp. 403–408 (1989).

[27] Y. ARACHI, H. SAKAI, O. YAMAMOTO, Y. TAKEDA and N. IMANISHAI. *Electrical conductivity of the ZrO_2–Ln_2O_3 (Ln= lanthanides) system*. Solid State Ionics, Vol. 7(1), pp. 133–139 (1999).

[28] J. LUO, D. P. ALMOND and R. STEVENS. *Ionic Mobilities and Association Energies from an Analysis of Electrical Impedance of ZrO_2–Y_2O_3 Alloys*. J. Am. Ceram. Soc., Vol. 83(7), pp. 1703–1708 (2000).

[29] W. BAUKAL. *Über die Kinetik der Alterung eines ZrO_2-Festelektrolyten in Abhängigkeit vom Sauerstoff-Partialdruck*. Electrochim. Acta, Vol. 14(11), pp. 1071–1080 (1969).

[30] F. T. CIACCHI, S. P. S. BADWAL and J. DRENNAN. *The System Y_2O_3–Sc_2O_3–ZrO_2: Phase Characterisation by XRD, TEM and Optical Microscopy*. J. Eur. Ceram. Soc., Vol. 7(3), pp. 185–195 (1991).

[31] F. T. CIACCHI and S. P. S. BADWAL. *The System Y_2O_3–Sc_2O_3–ZrO_2: Phase Stability and Ionic Conductivity Studies*. J. Eur. Ceram. Soc., Vol. 7(3), pp. 197–206 (1991).

[32] J. KONDOH, S. KIKUCHI, Y. TOMII and Y. ITO. *Effect of Aging on Yttria-Stabilized Zirconia: II. A Study of the Effect of the Microstructure on Conductivity*. J. Electrochem. Soc., Vol. 145(5), pp. 1536–1550 (1998).

[33] J. KONDOH, S. KIKUCHI, Y. TOMII and Y. ITO. *Effect of Aging on Yttria-Stabilized Zirconia: III. A Study of the Effect of Local Structures on Conductivity.* J. Electrochem. Soc., Vol. 145(5), pp. 1550–1560 (1998).

[34] J. KONDOH, S. KIKUCHI, Y. TOMII and Y. ITO. *Aging and composition dependence of electron diffraction patterns in Y_2O_3-stabilized ZrO_2: Relationship between crystal structure and conductivity.* Physica A, Vol. 262(1–2), pp. 177–189 (1999).

[35] B. BUTZ, R. SCHNEIDER, D. GERTHSEN, M. SCHOWALTER and A. ROSENAUER. *Decomposition of 8.5ámol.% Y_2O_3-doped zirconia and its contribution to the degradation of ionic conductivity.* Acta Materialia, Vol. 57(18), pp. 5480–5490 (2009).

[36] H. TULLER. *Solid State Electrochemical Systems–Opportunities for Nanofabricated or Nanostructured Materials.* Journal of Electroceramics, Vol. 1(3), pp. 211–218 (1997).

[37] H. L. TULLER. *Ionic conduction in nanocrystalline materials.* Solid State Ionics, Vol. 131(1), pp. 143–157 (2000).

[38] J. MAIER. *Thermodynamic aspects and morphology of nano-structured ion conductors. Aspects of nano-ionics. Part I.* Solid State Ionics, Vol. 154–155, pp. 291–301 (2002).

[39] J. MAIER. *Defect chemistry and ion transport in nanostructured materials. Aspects of nanoionics. Part II.* Solid State Ionics, Vol. 157(1-4), pp. 327–334 (2003).

[40] J. MAIER. *Nano-sized mixed conductors. Aspects of nano-ionics. Part III.* Solid State Ionics, Vol. 148(3–4), pp. 367–374 (2002).

[41] J. MAIER. *Nano-Ionics: Trivial and Non-Trivial Size Effects on Ion Conduction in Solids.* Z. Phys. Chem., Vol. 217(4), pp. 415–436 (2003).

[42] J. MAIER. *Ionic transport in nano-sized systems.* Solid State Ionics, Vol. 175(1–4), pp. 7–12 (2004).

[43] J. MAIER. *Transport in electroceramics: micro- and nano-structural aspects.* J. Eur. Ceram. Soc., Vol. 24(6), pp. 1251–1257 (2004).

[44] R. A. MILLER, J. L. SMIALEK and R. G. GARLICK. *Phase stability in plasma-sprayed, partially stabilized zirconia-yttria.* A. HEUER and L. HOBBS

(Eds.), *Advances in Ceramics: Science and Technology of Zirconia I*, Vol. 3, (pp. 241–253) (American Ceramics Society, Colombus, Ohio, 1980).

[45] P. B. ONAJI and J. K. COCHRAN. *Yttria stabilized zirconia thin films by multilayer electron-beam deposition.* Mater. Chem. Phys., Vol. 15(1), pp. 27–36 (1986).

[46] C. C. CHEN, M. M. NASRALLAH and H. U. ANDERSON. *Synthesis and characterization of YSZ thin-film electrolytes.* Solid State Ionics, Vol. 70–71, pp. 101–108 (1994).

[47] P. MONDAL and H. HAHN. *Investigation of the Complex Conductivity of Nanocrystalline Y_2O_3-Stabilized Zirconia.* Ber. Bunsenges. Phys. Chem., Vol. 101(11), pp. 1765–1768 (1997).

[48] P. CHARPENTIER, P. FRAGNAUD, D. SCHLEICH, Y. DENOS and E. GEHAIN. *Preparation of Thin Film SOFCs Working at Reduced Temperature.* Ionics, Vol. 4(1), pp. 118–123 (1998).

[49] I. KOSACKI, B. GORMAN and H. ANDERSON. *Microstructure and electrical conductivity in nanocrystalline oxide thin films.* T. RAMANARAYANAN (Ed.), *Ionic and Mixed Conducting Ceramics Vol. III*, (pp. 631–642) (Electrochemical Society, Pennington, 1998).

[50] P. MONDAL, A. KLEIN, W. JAEGERMANN and H. HAHN. *Enhanced specific grain boundary conductivity in nanocrystalline Y_2O_3-stabilized zirconia.* Solid State Ionics, Vol. 118(3), pp. 331–339 (1999).

[51] C. XIA, S. ZHA, W. YANG, R. PENG, D. PENG and G. MENG. *Preparation of yttria stabilized zirconia membranes on porous substrates by a dip-coating process.* Solid State Ionics, Vol. 133(3–4), pp. 287–294 (2000).

[52] I. KOSACKI, T. SUZUKI, V. PETROVSKY and H. U. ANDERSON. *Electrical conductivity of nanocrystalline ceria and zirconia thin films.* Solid State Ionics, Vol. 136–137, pp. 1225–1233 (2000).

[53] G. SOYEZ, J. A. EASTMAN, L. J. THOMPSON, G.-R. BAI, P. M. BALDO, A. W. MCCORMICK, R. J. DIMELFI, A. A. ELMUSTAFA, M. F. TAMBWE and D. S. STONE. *Grain-size-dependent thermal conductivity of nanocrystalline yttria-stabilized zirconia films grown by metal-organic chemical vapor deposition.* Appl. Phys. Lett., Vol. 77(8), pp. 1155–1157 (2000).

[54] C. WANG, W. L. WORRELL, S. PARK, J. M. VOHS and R. J. GORTE. *Fabrication and Performance of Thin-Film YSZ Solid Oxide Fuel Cells.* J. Electrochem. Soc., Vol. 148(8), pp. A864–A868 (2001).

[55] G. KNÖNER, K. REIMANN, R. RÖWER, U. SÖDERVALL and H.-E. SCHAEFER. *Enhanced oxygen diffusivity in interfaces of nanocrystalline $ZrO_2 \cdot Y_2O_3$.* Proceedings of the National Academy of Science, Vol. 100(7), pp. 3870–3873 (2003).

[56] V. PETROVSKY, T. SUZUKI, P. JASINSKI, T. PETROVSKY and H. U. ANDERSON. *Low-Temperature Processing of Thin-Film Electrolyte for Electrochemical Devices.* Electrochem. Solid-State Lett., Vol. 7(6), pp. A138–A139 (2004).

[57] I. KOSACKI, C. M. ROULEAU, P. F. BECHER, J. BENTLEY and D. H. LOWNDES. *Nanoscale effects on the ionic conductivity in highly textured YSZ thin films.* Solid State Ionics, Vol. 176(13–14), pp. 1319–1326 (2005).

[58] D. PEREDNIS, O. WILHELM, S. E. PRATSINIS and L. J. GAUCKLER. *Morphology and deposition of thin yttria-stabilized zirconia films using spray pyrolysis.* Thin Solid Films, Vol. 474(1–2), pp. 84–95 (2005).

[59] Y. PAN, J. H. ZHU, M. Z. HU and E. A. PAYZANT. *Processing of YSZ thin films on dense and porous substrates.* Surf. Coat. Tech., Vol. 200(5–6), pp. 1242–1247 (2005).

[60] N. H. MENZLER, R. HANSCH, M. GAUDON, H.-P. BUCHKREMER and D. STÖVER. *Preparation of solid oxide fuel cell electrolytes via sol-gel-route.* J. F. NIE and M. BARNETT (Eds.), Materials Forum Proceedings of the 3rd International Conference on Adcanced Materials Processing, Vol. 29, (pp. 403–407) (Melbourne, Australia, 2005).

[61] X. XU, C. XIA, S. HUANG and D. PENG. *YSZ thin films deposited by spin-coating for IT-SOFCs.* Ceram. Int., Vol. 31(8), pp. 1061–1064 (2005).

[62] O. WILHELM, S. E. PRATSINIS, D. PEREDNIS and L. J. GAUCKLER. *Electrospray and pressurized spray deposition of yttria-stabilized zirconia films.* Thin Solid Films, Vol. 479(1–2), pp. 121–129 (2005).

[63] J. H. JOO and G. M. CHOI. *Electrical conductivity of YSZ film grown by pulsed laser deposition.* Solid State Ionics, Vol. 177(11–12), pp. 1053–1057 (2006).

[64] Y. PARK, P. SU, S. CHA, Y. SAITO and F. PRINZ. *Thin-film SOFCs using gastight YSZ thin films on nanoporous substrates.* J. Electrochem. Soc., Vol. 153(2), pp. A431–A436 (2006).

[65] Y.-Y. CHEN and W.-C. J. WEI. *Processing and characterization of ultra-thin yttria-stabilized zirconia (YSZ) electrolytic films for SOFC.* Solid State Ionics, Vol. 177(3-4), pp. 351–357 (2006).

[66] T. H. SHIN, J. H. YU, S. LEE, I. S. HAN, S. K. WOO, B. K. JANG and S.-H. HYUN. *Preparation of YSZ Electrolyte for SOFC by Electron Beam PVD.* Key Eng. Mater., Vol. 317–318, pp. 913–916 (2006).

[67] P. BRIOIS, F. LAPOSTOLLE, V. DEMANGE, E. DJURADO and A. BILLARD. *Structural investigations of YSZ coatings prepared by DC magnetron sputtering.* Surf. Coat. Tech., Vol. 201(12), pp. 6012–6018 (2007).

[68] K. RODRIGO, J. KNUDSEN, N. PRYDS, J. SCHOU and S. LINDEROTH. *Characterization of yttria-stabilized zirconia thin films grown by pulsed laser deposition (PLD) on various substrates.* App. Surf. Sci., Vol. 254(4), pp. 1338–1342 (2007).

[69] S. GARCÍA-MARTÍN, D. P. FAGG and J. T. IRVINE. *Characterization of Diffuse Scattering in Yttria-Stabilized Zirconia by Electron Diffraction and High-Resolution Transmission Electron Microscopy.* Chemistry of Materials, Vol. 20(18), pp. 5933–5938 (2008).

[70] X.-J. NING, C.-X. LI, C.-J. LI and G.-J. YANG. *Modification of microstructure and electrical conductivity of plasma-sprayed YSZ deposit through post-densification process.* Mater. Sci. Eng. A, Vol. 428(1–2), pp. 98–105 (2006).

[71] J. SEYDEL. *Nanokristallines Zirkondioxid für Hochtemperatur-Brennstoffzellen.* PhD thesis, TU Darmstadt. URL http://tuprints.ulb.tu-darmstadt.de/462/ (2004).

[72] R. RAMAMOORTHY, D. SUNDARARAMAN and S. RAMASAMY. *Ionic conductivity studies of ultrafine-grained yttria stabilized zirconia polymorphs.* Solid State Ionics, Vol. 123(1), pp. 271–278 (1999).

[73] H. G. SCOTT. *Phase relationships in the zirconia-yttria system.* J. Mater. Sci., Vol. 10(9), pp. 1527–1535 (1975).

[74] O. RUFF and F. EBERT. *Beiträge zur Keramik hochfeuerfeste Stoffe. I. Die Formen des Zirkondioxyds.* Z. Anorg. Allg. Chem., Vol. 180(1), pp. 19–41 (1929).

[75] J. BÖHM. *Über das Verglimmen einiger Metalloxyde.* Z. Anorg. Allg. Chem., Vol. 149(1), pp. 217–222 (1925).

[76] J. D. MCCULLOUGH and K. N. TRUEBLOOD. *The crystal Structure of Baddeleyite (Monoclinic ZrO_2)*. Acta Crystallogr., Vol. 12(7), pp. 507–511 (1959).

[77] D. K. SMITH and H. W. NEWKIRK. *The Crystal Structure of Baddeleyite (Monoclinic ZrO_2) and its Relation to the Polymorphism of ZrO_2*. Acta Crystallogr., Vol. 18(6), pp. 983–991 (1965).

[78] R. E. HANN, P. R. SUITCH and J. L. PENTECOST. *Monoclinic Crystal Structures of ZrO_2 and HfO_2 Refined from X-ray Powder Diffraction Data*. J. Am. Ceram. Soc., Vol. 68(10), pp. C285–C286 (1985).

[79] G. TEUFER. *The crystal structure of tetragonal ZrO_2*. Acta Crystallogr., Vol. 15, p. 1187 (1962).

[80] R. N. PATIL and E. C. SUBBARAO. *Axial Thermal Expansion of ZrO_2 and HfO_2 in the Range Room Temperature to 1400 °C*. J. Appl. Crystallogr., Vol. 2(6), pp. 281–288 (1969).

[81] R. N. PATIL and E. C. SUBBARAO. *Monoclinic–Tetragonal Phase Transition in Zirconia: Mechanism, Pretransformation and Coexistence*. Acta Crystallogr. A, Vol. 26(5), pp. 535–542 (1970).

[82] M. RÜHLE and A. H. HEUER. *Phase Transformations in ZrO_2-Containing Ceramics: II, The Martensitic Reaction in t-ZrO_2*. M. RÜHLE, N. CLAUSSEN and A. H. HEUER (Eds.), *Advances in Ceramics: Science and Technology of Zirconia II*, Vol. 12, (pp. 14–32) (American Ceramics Society, Columbus, Ohio, 1984).

[83] D. R. CLARKE and A. ARORA. *Acoustic Emission Characterization of the Tetragonal–Monoclinic Phase Transformation in Zirconia*. M. RÜHLE, N. CLAUSSEN and A. H. HEUER (Eds.), *Advances in Ceramics: Science and Technology of Zirconia II*, Vol. 12, (pp. 54–63) (American Ceramics Society, Columbus, Ohio, 1984).

[84] C. J. HOWARD, R. J. HILL and B. E. REICHERT. *Structures of the ZrO_2 Polymorphs at Room Temperature by High-Resolution Neutron Powder Diffraction*. Acta Crystallogr. B, Vol. 44(2), pp. 116–120 (1988).

[85] F. FREY, H. BOYSEN and T. VOGT. *Neutron Powder Investigation of the Monoclinic to Tetragonal Phase Transformation in Undoped Zirconia*. Acta Crystallogr. B, Vol. 46(6), pp. 724–730 (1990).

[86] E. C. SUBBARAO, H. S. MAITI and K. K. SRIVASTAVA. *Martensitic Transformation in Zirconia*. Phys. Status Solidi A, Vol. 21(1), pp. 9–40 (1974).

[87] G. BANSAL and A. H. HEUER. *On a martensitic phase transformation in zirconia (ZrO_2)-I. metallographic evidence.* Acta Metall., Vol. 20(11), pp. 1281–1289 (1972).

[88] G. BANSAL and A. H. HEUER. *On a martensitic phase transformation in zirconia (ZrO_2)-II. crystallographic aspects.* Acta Metall., Vol. 22(4), pp. 409–417 (1974).

[89] M. YASHIMA, M. KAKIHANA and M. YOSHIMURAR. *Meta–stable phase diagrams in the zirconia-containing systems utilized in solid-oxide fuell cell application.* Solid State Ionics, Vol. 86–88(2), pp. 1131–1149 (1996).

[90] D. K. SMITH and C. F. CLINE. *Verification of Existence of Cubic Zirconia at High Temperature.* J. Am. Ceram. Soc., Vol. 45(5), pp. 249–250 (1962).

[91] A. ROUANET. C. R. Acad. Sci. Paris Series C, Vol. 267, pp. 395–397 (1968).

[92] M. CHEN, B. HALLSTEDT and L. J. GAUCKLER. *Thermodynamic modeling of the $ZrO_2-YO_{1.5}$ system.* Solid State Ionics, Vol. 170(3–4), pp. 255–274 (2004).

[93] R. C. GARVIE. *The Occurrence of Metastable Tetragonal Zirconia as a Crystallite Size Effect.* J. Phys. Chem., Vol. 69(4), pp. 1238–1243 (1965).

[94] M. G. PATON and E. N. MASLEN. *A Refinement of the Crystal Structure of Yttria.* Acta Crystallogr., Vol. 19(3), pp. 307–310 (1965).

[95] I. WARSHAW and R. ROY. *Polymorphism of the Rare Earth Sesquioxides.* J. Phys. Chem., Vol. 65, pp. 2048Ű-2051 (1961).

[96] A. BARTOS, K. P. LIEB, M. UHRMACHER and D. WIARDA. *Refinement of Atomic Positions in Bixbyite Oxides using Perturbed Angular Correlation Spectroscopy.* Acta Crystallogr. B, Vol. 49(2), pp. 165–169 (1993).

[97] P. DUWEZ, F. H. B. JR. and F. ODELL. *The Zirconia-Yttria System.* J. Electrochem. Soc., Vol. 98(9), pp. 356–362 (1951).

[98] F. FU-KANG, A. K. KUZNETSOV and E. K. KELER. *Phase Relationships in the $Y_2O_3-ZrO_2$ System. Part 2. Solid Solutions.* Izv. Akad. Nauk SSSR, Ser. Khim., translation: Otdelenie Khimicheskikh Nauk, Bulletin of the Academy of Sciences, Division of Chemical Science, Vol. 4, pp. 542–549 (1963).

[99] C. PASCUAL and P. DURÁN. *Subsolidus Phase Equilibria and Ordering of the System ZrO_2-Y_2O_3.* J. Am. Ceram. Soc., Vol. 66(1), pp. 23–27 (1983).

[100] R. Ruh, K. S. Mazdiyasni, P. G. Valentine and H. O. Bielstein. *Phase Relations in the System ZrO_2-Y_2O_3 at Low Y_2O_3 Contents*. J. Am. Ceram. Soc., Vol. 67(9), pp. C190–C192 (1984).

[101] G. S. Corman and V. S. Stubican. *Phase Equilibria and Ionic Conductivity in the System ZrO_2-Yb_2O_3-Y_2O_3*. J. Am. Ceram. Soc., Vol. 68(4), pp. 174–181 (1985).

[102] H. Suto, T. Sakuma and N. Yoshikawa. *Discussion on the Phase Diagram of Y_2O_3-Partially Stabilized Zirconia and Interpretation of the Structures*. T. Jpn. I. Met., Vol. 28(8), pp. 623–630 (1987).

[103] M. Yoshimura. *Phase Stability of Zirconia*. Ceramic Bulletin, Vol. 67(12), pp. 1950–1955 (1988).

[104] V. S. Stubican. *Phase Equilibria and Metastabilities in the Systems ZrO_2-MgO, ZrO_2-CaO, and ZrO_2-Y_2O_3*. S. Somiya, N. Yamamoto and H. Yanagida (Eds.), *Advances in Ceramics: Science and Technology of Zirconia III*, Vol. 24, (pp. 71–82) (American Ceramics Society, Westerville, Ohio, 1988).

[105] N. R. Rebollo, O. Fabrichnaya and C. G. Levi. *Phase stability of Y + Gd co-doped zirconia*. Z. Metallkd., Vol. 94(3), pp. 163–170 (2003).

[106] Y. Sakka, Y. Oishi and K. Ando. *Zr–Hf interdiffusion in polycrystalline Y_2O_3-$(Zr+Hf)O_2$*. J. Mater. Sci., Vol. 17(11), pp. 3101–3105 (1982).

[107] M. Kilo, G. Borchardt, B. Lesage, O. Kaitasov, S. Weber and S. Scherrer. *Cation transport in yttria stabilized cubic zirconia: ^{96}Zr tracer diffusion in $(Zr_xY_{1-x})O_{2-x/2}$ single crystals with $0.15 \leqslant x \leqslant 0.48$*. J. Eur. Ceram. Soc., Vol. 20(12), pp. 2069–2077 (2000).

[108] M. Kilo, M. Weller, G. Borchardt, B. Damson, S. Weber and S. Scherrer. *Cation Mobility in Y_2O_3- and CaO-Stabilised ZrO_2 Studied by Tracer Diffusion and Mechanical Spectroscopy*. Defect and Diffusions Forum, Vol. 194–199, pp. 1039–1044 (2001).

[109] M. Kilo, M. A. Taylor, C. Argirusis, G. Borchardt, B. Lesage, S. Weber, S. Scherrer, H. Scherrer, M. Schroeder and M. Martin. *Cation self-diffusion of ^{44}Ca, ^{88}Y, and ^{96}Zr in single-crystalline calcia- and yttria-doped zirconia*. J. Appl. Phys., Vol. 94(12), pp. 7547–7552 (2003).

[110] M. KILO, M. A. TAYLOR, C. ARGIRUSIS, G. BORCHARDT, S. WEBER, H. SCHERRER and R. A. JACKSON. *Lanthanide transport in stabilized zirconias: Interrelation between ionic radius and diffusion coefficient.* J. Chem. Phys., Vol. 121(11), pp. 5482–5487 (2004).

[111] S. SWAROOP, M. KILO, C. ARGIRUSIS, G. BORCHARDT and A. H. CHOKSHI. *Lattice and grain boundary diffusion of cations in 3YTZ analyzed using SIMS.* Acta Mater., Vol. 53(19), pp. 4975–4985 (2005).

[112] V. LANTERI, A. H. HEUER and T. E. MITCHELL. *Tetragonal Phase in the System ZrO_2-Y_2O_3.* M. RÜHLE, N. CLAUSSEN and A. H. HEUER (Eds.), Advances in Ceramics: Science and Technology of Zirconia II, Vol. 12, (pp. 118–130) (American Ceramics Society, Columbus, Ohio, 1984).

[113] A. H. HEUER and M. RÜHLE. *Phase Transformations in ZrO_2-Containing Ceramics: I, The Instability of c-ZrO_2 and the Resulting Diffusion-Controlled Reactions.* M. RÜHLE, N. CLAUSSEN and A. H. HEUER (Eds.), Advances in Ceramics: Science and Technology of Zirconia II, Vol. 12, (pp. 1–13) (American Ceramics Society, Columbus, Ohio, 1984).

[114] R. CHAIM, M. RÜHLE and A. H. HEUER. *Microstructural Evolution in a ZrO_2-12 Wt% Y_2O_3 Ceramic.* J. Am. Ceram. Soc., Vol. 68(8), pp. 427–431 (1985).

[115] M. RÜHLE, L. T. MA, W. WUNDERLICH and A. G. EVANS. *TEM studies on phases and phase stabilities of zirconia ceramics.* Physica B, Vol. 150, pp. 86–98 (1988).

[116] Y. ZHOU, Q. L. GE and T. C. LEI. *Diffusional cubic-to-tetragonal phase transformation and microstructural evolution in ZrO_2-Y_2O_3 ceramics.* J. Mater. Sci., Vol. 26(16), pp. 4416–4467 (1991).

[117] T. SAKUMA, Y. YOSHIZAWA and H. SUTO. *The metastable two-phase region in the zirconia-rich part of the ZrO_2-Y_2O_3 system.* J. Mater. Sci., Vol. 21(4), pp. 1436–1440 (1986).

[118] M. DOI and T. MIYAZAKI. *On the spinodal decomposition in zirconia-yttria (ZrO_2-Y_2O_3) alloys.* Philos. Mag. B, Vol. 68(3), pp. 305–315 (1993).

[119] J. KATAMURA, N. SHIBATA, Y. IKUHARA and T. SAKUMA. *Transmission electron microscopy–energy-dispersive X-ray spectroscopy analysis of the modulated structure in ZrO_2-6 mol% Y_2O_3 alloy.* Philos. Mag. Lett., Vol. 78(1), pp. 45–49 (1998).

[120] N. SHIBATA, J. KATAMURA, A. KUWABARA, Y. IKUHARA and T. SAKUMA. *The instability and resulting phase transition of cubic zirconia.* Mater. Sci. Eng. A, Vol. 312(1–2), pp. 90–98 (2001).

[121] R. F. GELLER and P. J. YAVORSKY. *Effects of some oxide additions on the thermal length changes of zirconia.* J. Res. Nat. Bur. Stand., Vol. 35, pp. 87–110 (1945).

[122] V. SRIKANTH and E. C. SUBBARAO. *Acoustic emission study of phase relations in low-Y_2O_3 portion of ZrO_2-Y_2O_3 system.* J. Mater. Sci., Vol. 29(12), pp. 3363–3371 (1994).

[123] S. A. DEGTYAREV and G. F. VORININ. *Solution of Incorrect Problems in the Thermodynamics of Phase Equilibria. I. The System ZrO_2-Y_2O_3.* Russ. J. Phys. Chem., Vol. 61(3), pp. 317–320 (1987).

[124] S. A. DEGTYAREV and G. F. VORONIN. *Solution of ill-posed problems in thermodynamics of phase equilibria. The ZrO_2-Y_2O_3 system.* Calphad, Vol. 12(1), pp. 73–82 (1988).

[125] K. KOBAYASHI, H. KUWAJIMA and T. MASAKI. *Phase change and mechanical properties of ZrO_2-Y_2O_3 solid electrolyte after ageing.* Solid State Ionics, Vol. 3–4, pp. 489–493 (1981).

[126] F. F. LANGE. *Transformation toughening. Part 3 Experimental observations in the ZrO_2-Y_2O_3 system.* J. Mater. Sci., Vol. 17(1), pp. 240–246 (1982).

[127] A. H. HEUER, N. CLAUSSEN, W. M. KRIVEN and M. RÜHLE. *Stability of Tetragonal ZrO_2 Particles in Ceramic Matrices.* J. Am. Ceram. Soc., Vol. 65(12), pp. 642–650 (1982).

[128] A. H. HEUER and M. RÜHLE. *On the nucleation of the martensitic transformation in zirconia (ZrO_2).* Acta Metall., Vol. 33(12), pp. 2101–2112 (1985).

[129] A. SURESH, M. J. MAYO and W. D. PORTER. *Thermodynamics of the tetragonal-to-monoclinic phase transformation in fine and nanocrystalline yttria-stabilized zirconia powders.* J. Mater. Res., Vol. 18(12), pp. 2912–2921 (2003).

[130] M. J. MAYO, A. SURESH and W. D. PORTER. *Thermodynamics for nanosystems: grain and particle-size dependent phase diagrams.* Reviews on Advanced Materials Science, Vol. 5(2), pp. 100–109 (2003).

[131] R. H. J. HANNINK. *Growth morphology of the tetragonal phase in partially stabilized zirconia.* J. Mater. Sci., Vol. 13(11), pp. 2487–2496 (1978).

[132] D. MICHEL, L. MAZEROLLES and R. PORTIER. *Electron microscopy observation of the domain boundaries generated by the cubic → tetragonal transition of stabilized zirconia.* R. METSELAAR, H. J. M. HEIJLIGERS and J. SCHOOMAN (Eds.), *Studies in Inorganic Chemistry, Proceedings of the Second European Conference*, Vol. 3, (pp. 809–812) (Veldhoven, Netherlands, 1982).

[133] M. SUGIYAMA and H. KUBO. *Microstructure of the Cubic and Tetragonal Phases in a ZrO_2-Y_2O_3 Ceramic System.* M. RÜHLE, N. CLAUSSEN and A. H. HEUER (Eds.), *Advances in Ceramics: Science and Technology of Zirconia II*, Vol. 12, (pp. 965–973) (American Ceramics Society, Columbus, Ohio, 1984).

[134] C. A. ANDERSSON, J. G. JR. and T. K. GUPTA. *Diffusionless Transformations in Zirconia Alloys.* M. RÜHLE, N. CLAUSSEN and A. H. HEUER (Eds.), *Advances in Ceramics: Science and Technology of Zirconia II*, Vol. 12, (pp. 78–85) (American Ceramics Society, Columbus, Ohio, 1984).

[135] V. LANTERI, R. CHAIM and A. H. HEUER. *On the Microstructures Resulting from the Diffusionless Cubic ⟶ Tetragonal Transformation in ZrO_2-Y_2O_3 Alloys.* J. Am. Ceram. Soc., Vol. 69(10), pp. C-258–C-261 (1986).

[136] A. H. HEUER, R. CHAIM and V. LANTERI. *The displacive cubic → tetragonal transformation in ZrO_2 alloys.* Acta Metall., Vol. 35(3), pp. 661–666 (1987).

[137] T. NOMA, M. YOSHIMURA, S. SOMIYA, M. KATO, M. S. YANAGISAWA and H. SETO. *Formation of diffusionlessly transformed tetragonal phases in rapid quenching of ZrO_2-Y_2O_3 melts.* J. Mater. Sci., Vol. 23(8), pp. 2689–2692 (1988).

[138] T. NOMA, M. YOSHIMURA and S. SOMIYA. *Stability of Diffusionlessly Transformed Tetragonal Phases in Rapidly Quenched ZrO_2-Y_2O_3.* S. SOMIYA, N. YAMAMOTO and H. YANAGIDA (Eds.), *Advances in Ceramics: Science and Technology of Zirconia III*, Vol. 24, (pp. 377–384) (American Ceramics Society, Westerville, Ohio, 1988).

[139] M. YOSHIMURA, M. YASHIMA, T. NOMA and S. SOMIYA. *Formation of diffusionlessly transformed tetragonal phases by rapid quenching of melts in ZrO_2-$RO_{1.5}$ systems (R = rare earths).* J. Mater. Sci., Vol. 25(4), pp. 2011–2016 (1990).

[140] L. LELAIT and S. ALPÉRINE. *TEM investigations of high toughness non-equilibrium phases in the ZrO_2-Y_2O_3 system.* Scripta Metall. Mater., Vol. 25(8), pp. 1815–1820 (1991).

[141] Y. ZHOU, T.-C. LEI and T. SAKUMA. *Diffusionless Cubic-to-Tetragonal Phase Transition and Microstructural Evolution in Sintered Zirconia–Yttria Ceramics.* J. Am. Ceram. Soc., Vol. 74(3), pp. 633–640 (1991).

[142] M. YASHIMA, N. ISHIZAWA, H. FUJIMORI, M. KAKIHANA and M. YOSHIMURA. *In situ observation of the diffusionless tetragonal $\rightleftarrows$ cubic phase transition and metastable-stable phase diagram in the ZrO_2-$ScO_{1.5}$ system.* Eur. J. Solid State Inorg. Chem., Vol. 32(7–8), pp. 761–770 (1995).

[143] D. FAN and L.-Q. CHEN. *Computer Simulation of Twin Formation during the Displacive $c \longrightarrow t'$ Phase Transformation in the Zirconia–Yttria System.* J. Am. Ceram. Soc., Vol. 78(3), pp. 769–773 (1995).

[144] I. R. GIBSON and J. T. S. IRVINE. *Qualitative X-ray Diffraction Analysis of Metastable Tetragonal (t') zirconia.* J. Am. Ceram. Soc., Vol. 84(3), pp. 615–618 (2001).

[145] J. LEFÈVRE. *Contribution à l'étude de différentes modifications structurales des phases de type fluorine dans les systèmes à base de zircone ou d'oxyde de hafnium.* Ann. Chim., Vol. 8(1–2), pp. 117–149 (1963).

[146] M. YASHIMA, N. ISHIZAWA and M. YOSHIMURA. *High Temperature X-Ray Diffraction Study on Cubic-Tetragonal Phase Transition in the ZrO_2-$RO_{1.5}$ Systems (R: Rare Earths).* S. P. BADWAL, M. J. BANNISTER and R. H. J. HANNINK (Eds.), Science and Technology of Zirconia V, (pp. 125–135) (Techomic Publishing Co., Lancaster, 1993).

[147] M. YASHIMA, S. SASAKI, M. KAKIHANA, Y. YAMAGUCHI, H. ARASHI and M. YOSHIMURA. *Oxygen-Induced Structural Change of the Tetragonal Phase Around the Tetragonal–Cubic Phase Boundary in ZrO_2-$YO_{1.5}$ Solid Solutions.* Acta Crystallogr. B, Vol. 50(6), pp. 663–672 (1994).

[148] M. YASHIMA, K. OHTAKE, M. KAKIHANA, H. ARASHI and M. YOSHIMURA. *Determination of tetragonal–cubic phase boundary of $Zr_{1-X}R_X O_{2-X/2}$ (R=Nd, Sm, Y, Er and Yb) by Raman scattering.* J. Phys. Chem. Solids, Vol. 57(1), pp. 17–24 (1996).

[149] Y. DU, Z. JIN and P. HUANG. *Thermodynamic Assessment of the ZrO_2-$YO_{1.5}$ System.* J. Am. Ceram. Soc., Vol. 74(7), pp. 1569–1577 (1991).

[150] M. HILLERT and T. SAKUMA. *Thermodynamic modeling of the $c \longrightarrow t$ transformation in ZrO_2 alloys.* Acta Metall. Mater., Vol. 39(6), pp. 1111–1115 (1991).

[151] L. LI, O. VAN DER BIEST, P. L. WANG, J. VLEUGELS, W. W. CHEN and S. G. HUANG. *Estimation of the phase diagram for the ZrO_2-Y_2O_3-CeO_2 system*. J. Eur. Ceram. Soc., Vol. 21(16), pp. 2903–2910 (2001).

[152] O. FABRICHNAYA and F. ALDINGER. *Assessment of thermodynamic parameters in the system ZrO_2-Y_2O_3-Al_2O_3*. Z. Metallkd., Vol. 95(1), pp. 27–39 (2004).

[153] N. S. JACOBSON, Z.-K. LIU, L. KAUFMAN and F. ZHANG. *Thermodynamic Modeling of the $YO_{1.5}$-ZrO_2 system*. J. Am. Ceram. Soc., Vol. 87(8), pp. 1559–1566 (2004).

[154] M. CHEN, B. HALLSTEDT and L. J. GAUCKLER. *Thermodynamic modeling of phase equilibria in the Mn–Y–Zr–O system*. Solid State Ionics, Vol. 176(15–16), pp. 1457–1464 (2005).

[155] M. RÜHLE, N. CLAUSSEN and A. H. HEUER. *Microstructural Studies of Y_2O_3-Containing Tetragonal ZrO_2 Polycrystals (Y-TZP)*. M. RÜHLE, N. CLAUSSEN and A. H. HEUER (Eds.), Advances in Ceramics: Science and Technology of Zirconia II, Vol. 12, (pp. 352–370) (American Ceramics Society, Columbus, Ohio, 1984).

[156] H. SCHUBERT, N. CLAUSSEN and M. RÜHLE. *Preparation of Y_2O_3-Stabilized Tetragonal ZrO_2 Polycrystals (Y-TZP) from Different Powders*. M. RÜHLE, N. CLAUSSEN and A. H. HEUER (Eds.), Advances in Ceramics: Science and Technology of Zirconia II, Vol. 12, (pp. 766–773) (American Ceramics Society, Columbus, Ohio, 1984).

[157] N. M. BEEKMANS and L. HEYNE. *Correlation between impedance, microstructure and composition of calcia-stabilized zirconia*. Electrochim. Acta, Vol. 21(4), pp. 303–310 (1976).

[158] S. P. S. BADWAL and J. DRENNAN. *Interfaces in zirconia based electrochemical systems and their influence on electrical properties*. J. NOWOTNY (Ed.), Science of Ceramic Interfaces II, Vol. 1, (pp. 71–111) (Elsevier Science Pub Co, Elsevier, Amsterdam, 1994).

[159] S. P. S. BADWAL and M. V. SWAIN. *ZrO_2-Y_2O_3: electrical conductivity of some fully and partially stabilized single grains*. J. Mater. Sci. Lett., Vol. 4(4), pp. 487–489 (1985).

[160] M. AOKI, Y.-M. CHIANG, I. KOSACKI, L. J.-R. LEE, H. TULLER and Y. LIU. *Solute Segregation and Grain-Boundary Impedance in High-Purity Stabilized Zirconia*. J. Am. Ceram. Soc., Vol. 79(5), pp. 1169–1180 (1996).

[161] S. STEMMER, J. VLEUGELS and O. V. D. BIEST. *Grain Boundary Segregation in High-purity, Yttria-stabilized Tetragonal Zirconia Polycrystals (Y-TZP)*. J. Eur. Ceram. Soc., Vol. 18(11), pp. 1565–1570 (1998).

[162] S. P. S. BADWAL, F. T. CIACCHI, S. RAJENDRAN and J. DRENNAN. *An investigation of conductivity, microstructure and stability of electrolyte compositions in the system 9 mol% (Sc_2O_3–Y_2O_3)–ZrO_2(Al_2O_3)*. Solid State Ionics, Vol. 109(3–4), pp. 167–186 (1998).

[163] J.-H. LEE, T. MORI, J.-G. LI, T. IKEGAMI, M. KOMATSU and H. HANEDA. *Improvement of Grain-Boundary Conductivity of 8 mol% Yttria-Stabilized Zirconia by Precursor Scavenging of Siliceous Phase*. J. Electrochem. Soc., Vol. 147(7), pp. 2822–2829 (2000).

[164] R. P. INGEL and D. LEWISIII. *Lattice Parameters and Density for Y_2O_3-Stabilized ZrO_2*. J. Am. Ceram. Soc., Vol. 69(4), pp. 325–332 (1986).

[165] F. HANIC, M. HARTMANOVÁ, F. KUNDRACIK and E. E. LOMONOVA. *Stabilization and Properties of High Temperature Forms of Zirconia*. Solid State Phenom., Vol. 90–91, pp. 303–308 (2003).

[166] A. MÜLLER. *Mehrschicht-Anode für die Hochtemperatur-Brennstoffzelle (SOFC)*. PhD thesis, University of Karlsruhe (TH). URL http://digbib.ubka.uni-karlsruhe.de/volltexte/1000003380 (2004).

[167] C. PETERS, A. WEBER, B. BUTZ, D. GERTHSEN and E. IVERS-TIFFÉE. *Grain-Size Effects in YSZ Thin-Film Electrolytes*. J. Am. Ceram. Soc., Vol. 92(9), pp. 2017–2024 (2009).

[168] P. STADELMANN. *JEMS*. URL http://cimewww.epfl.ch/people/stadelmann/jemsWebSite/jems.html.

[169] *Fityk – free peak fitting software*. URL http://www.unipress.waw.pl/fityk/.

[170] *ImageJ – Image Processing and Analysis in Java*. URL http://rsbweb.nih.gov/ij/.

[171] G. W. LORIMER. *Quantitative X-ray microanalysis of thin specimens in the transmission electron microscope; a review*. Mineral. Mag., Vol. 51(359), pp. 49–60 (1987).

[172] J. H. HUBBELL. *Photon mass attenuation and energy-absorption coefficients*. The International Journal of Applied Radiation and Isotopes, Vol. 33(11), pp. 1269–1290 (1982).

[173] M. R. TALUKDER, S. BOSE and S. TAKAMURA. *Calculated electron impact K-shell ionization cross sections for atoms.* Int. J. Mass Spectrom., Vol. 269(1–2), pp. 118–130 (2008).

[174] M. O. KRAUSE. *Atomic Radiative and Radiationless Yields of K and L Shells.* J. Phys. Chem. Ref. Data, Vol. 8(2), pp. 307–324 (1979).

[175] O. ŞIMŞEK, M. ERTUGRUL, D. KARAGÖZ, G. BUDAK, A. KARABULUT, S. YILMAZ, O. DOĞAN, U. TURGUT, O. SÖĞÜT, R. POLAT and A. GÜROL. *K shell fluorescence yields for elements with $33 \leq Z \leq 53$ using 59.5 keV photons.* Radiat. Phys. Chem., Vol. 65(1), pp. 27–31 (2002).

[176] J. H. SCOFIELD. *Exchange corrections of K x-ray emission rates.* Phys. Rev. A, Vol. 9(3), pp. 1041–1049 (1974).

[177] T. MALIS, S. C. CHENG and R. F. EGERTON. *EELS Log-Ratio Technique for Specimen-Thickness Measurement in the TEM.* J. Electron. Micr. Tech., Vol. 8(2), pp. 193–200 (1988).

[178] K. IAKOUBOVSKII, K. MITSUISHI, Y. NAKAYAMA and K. FURUYA. *Mean free path of inelastic electron scattering in elemental solids and oxides using transmission electron microscopy: Atomic number dependent oscillatory behavior.* Phys. Rev. B, Vol. 77(10), pp. 104102-1–104102-7 (2008).

[179] R. EGERTON. *Electron Energy-Loss Spectroscopy in the Electron Microscope* (Plenum Press, New York, 1986).

[180] R. BRYDSON. *Electron Energy loss spectroscopy* (BIOS Scientific Publishers Ltd., Oxford, 2001).

[181] T. MALIS and J. TITCHMARSH. *Şk-factorŤ approach to EELS analysis. Electron Microscopy and Analysis*, (pp. 181–182) (Institute of Physics, Bristol, UK, 1986).

[182] R. EGERTON. *K-shell ionization cross-sections for use in microanalysis.* Ultramicroscopy, Vol. 4(2), pp. 169–179 (1979).

[183] R. D. LEAPMAN, P. REZ and D. F. MAYERS. *K, L, and M shell generalized oscillator strengths and ionization cross sections for fast electron collisions.* J. Chem. Phys., Vol. 72(2), pp. 1232–1243 (1980).

[184] P. REZ. *Cross-sections for energy loss spectrometry.* Ultramicroscopy, Vol. 9, pp. 283–288 (1982).

[185] C. AHN and P. REZ. *Inner shell edge profiles in electron energy loss spectroscopy.* Ultramicroscopy, Vol. 17(2), pp. 105–116 (1985).

[186] P. LÖBMANN, R. JAHN, S. SEIFERT and D. SPORN. *Inorganic Thin Films Prepared from Soluble Powders and Their Applications.* Journal of Sol-Gel Science and Technology, Vol. 19(1), pp. 473–477 (2000).

[187] P. C. LÖBMANN, W. GLAUBITT and D. SPORN. *Verfahren zur Abscheidung von Zirkonoxid-Schichten unter Verwendung von löslichen Pulvern.* European patent ep 1 084 992 a1, Fraunhofer-Institut für Silicatforschung ISC.

[188] R. KRÜGER, M. J. BOCKMEYER, A. DUTSCHKE and P. C. LÖBMANN. *Continuous Sol-Gel Coating of Ceramic Multifilaments: Evaluation of Fiber Bridging by Three-Point Bending Test.* J. Am. Ceram. Soc., Vol. 89(7b), pp. 2080–2088 (2006).

[189] C. PETERS, M. BOCKMEYER, R. KRÜGER, A. WEBER and E. IVERS-TIFFÉE. *Processing of Dense Nanocrystalline Zirconia Thin Films by Sol-Gel.* D. KUMAR, V. CRACIUN, M. ALEXE and K. K. SINGH (Eds.), *Current and Future Trends of Functional Oxide Films*, Vol. 928E (2006).

[190] J. LI, T. MALIS and S. DIONNE. *Recent advances in FIB-TEM specimen preparation techniques.* Mater. Charact., Vol. 57(1), pp. 64–70 (2006).

[191] B. BUTZ. *Transmissionselektronenmikroskopische Strukturuntersuchungen an gesintertem und gealtertem Y_2O_3-dotiertem ZrO_2.* Diploma thesis, University of Karlsruhe (2004).

[192] B. BUTZ, H. STÖRMER, D. GERTHSEN, M. BOCKMEYER, R. KRÜGER, E. IVERS-TIFFÉE and M. LUYSBERG. *Microstructure of Nanocrystalline Yttria-Doped Zirconia Thin Films Obtained by Sol–Gel Processing.* J. Am. Ceram. Soc., Vol. 91(7), pp. 2281–2289 (2008).

[193] T. SAKUMA and H. SOTO. *Cubic-Tetragonal Phase Separations in Y-PSZ.* S. SOMIYA, N. YAMAMOTO and H. YANAGIDA (Eds.), *Advances in Ceramics: Science and Technology of Zirconia III*, Vol. 24, (pp. 531–535) (American Ceramics Society, Westerville, Ohio, 1988).

[194] J. W. CAHN. *On spinodal decomposition.* Acta Metall., Vol. 9, pp. 795–801 (1965).

[195] W. KINGERY, H. BOWEN and D. UHLMANN. *Introduction to Ceramics.* Second ed. (John Wiley & Sons Ltd., 1976).

[196] A. N. VLASOV and M. V. PERFILIEV. *Ageing of ZrO_2-based solid electrolytes.* Solid State Ionics, Vol. 25(4), pp. 245–253 (1987).

[197] C. G. KONTOYANNIS and M. ORKOULA. *Quantitative determination of the cubic, tetragonal and monoclinic phases in partially stabilized zirconias by Raman spectroscopy.* J. Mater. Sci., Vol. 29(20), pp. 5316–5320 (1994).

[198] S. GARCÍA-MARTÍN, M. A. ALARIO-FRANCO, D. P. FAGG and J. T. S. IRVINE. *Evidence of three types of short range ordered fluorite structure in the $(1-x)Y_{0.15}Zr_{0.85}O_{1.93}-xY_{0.75}Nb_{0.25}O_{1.75}$ $(0 \leq x \leq 1)$ system.* J. Mater. Chem., Vol. 15(19), pp. 1903–1907 (2005).

[199] K. MILLER and F. LANGE. *Highly oriented thin films of cubic zirconia on sapphire through grain growth seeding.* J. Mater. Res., Vol. 6(11), pp. 2387–2392 (1991).

[200] B. P. GORMAN and H. U. ANDERSON. *Microstructure Development in Unsupported Thin Films.* J. Am. Ceram. Soc., Vol. 85(4), pp. 981–985 (2002).

[201] Q. ZHU and B. FAN. *Low temperature sintering of 8YSZ electrolyte film for intermediate temperature solid oxide fuel cells.* Solid State Ionics, Vol. 176(9-10), pp. 889–894 (2005).

[202] A. DÍAZ-PARRALEJO, R. CARUSO, A. L. ORTIZ and F. GUIBERTEAU. *Densification and porosity evaluation of ZrO_2-3 mol.% Y_2O_3 sol-gel thin films.* Thin Solid Films, Vol. 458(1-2), pp. 92–97 (2004).

[203] C. SAKURAI, T. FUKUI and M. OKUYAMA. *Preparation of Zirconia Coatings by Hydrolysis of Zirconium Alkoxide with Hydrogen Peroxide.* J. Am. Ceram. Soc., Vol. 76(4), pp. 1061–1064 (1993).

[204] F. C. M. WOUDENBERG, W. F. C. SAGER, J. E. ELSHOF and H. VERWEIJ. *Nanostructured Dense ZrO_2 Thin Films from Nanoparticles Obtained by Emulsion Precipitation.* J. Am. Ceram. Soc., Vol. 87(8), pp. 1430–1435 (2004).

[205] R. NEAGU, E. DJURADO, L. ORTEGA and T. PAGNIER. *ZrO_2-based thin films synthesized by electrostatic spray deposition: Effect of post-deposition thermal treatments.* Solid State Ionics, Vol. 177(17-18), pp. 1443–1449 (2006).

[206] M. N. RAHAMAN. *Sintering of Ceramics*, Ch. Grain growth and microstructure control, (pp. 105–176) (CRC Press, Taylor & Francis Group, Boca Raton, FL, 2008).

[207] J. L. M. RUPP, A. INFORTUNA and L. J. GAUCKLER. *Microstrain and self-limited grain growth in nanocrystalline ceria ceramics.* Acta Mater., Vol. 54(7), pp. 1721–1730 (2006).

[208] M. F. YAN. *Microstructural Control in the Processing of Electronic Ceramics.* Mater. Sci. Eng., Vol. 48(1), pp. 53–72 (1981).

Publications and Contributions to Conferences

Related publications

- B. BUTZ, P. KRUSE, H. STÖRMER, D. GERTHSEN, A. MÜLLER, A. WEBER and E. IVERS-TIFFÉE. *Correlation between microstructure and degradation in conductivity for cubic Y_2O_3-doped ZrO_2.* Solid State Ionics, Vol. 177(37–38), pp. 3275–3284 (2006).

- B. BUTZ, H. STÖRMER, D. GERTHSEN, R. KRÜGER, E. IVERS-TIFFÉE and M. LUYSBERG. *Microstructure of nanocrystalline Y-doped zirconia thin films obtained by sol-gel processing.* J. Am. Ceram. Soc., Vol. 91(7), pp. 2281–2289 (2008).

- S.J. LITZELMAN, R.A. DE SOUZA, B. BUTZ, D. GERTHSEN, M. MARTIN and H.L. TULLER. *Heterogeneously doped nanocrystalline ceria films by grain boundary diffusion: Impact on transport properties.* J. Electroceram., Vol. 22(4), pp. 405–415 (2009).

- B. BUTZ, R. SCHNEIDER, D. GERTHSEN, M. SCHOWALTER and A. ROSENAUER. *Decomposition of 8.5 mol.% Y_2O_3-doped zirconia and its contribution to the degradation of ionic conductivity.* Acta Mater., Vol. 57(18), pp. 5480–5490 (2009).

- C. PETERS, A. WEBER, B. BUTZ, D. GERTHSEN and E. IVERS-TIFFÉE. *Grain-Size Effects in YSZ Thin-Film Electrolytes.* J. Am. Ceram. Soc., Vol. 92(9), pp. 2017–2024 (2009).

Further Publications

- S. ROY, B. BUTZ and A. WANNER. *Damage Evolution and Anisotropy in Freeze Cast Metals/Ceramic Composites: an IN-SITU SEM Analysis.* Proceedings of 13th European Conference on Composite Materials (ECCM-13), paper No. 0303 (2008).

- S. ROY, B. BUTZ and A. WANNER. *In-situ study of damage evolution and domain level anisotropy in an innovative metal/ceramic composite exhibiting lamellar microstructure.* Acta Mater, submitted April (2009).

Contributions to Conferences

- B. BUTZ, H. STÖRMER, D. GERTHSEN, A. MÜLLER, and E. IVERS-TIFFÉE. *Correlation between microstructure and degradation of ionic conductivity in Y_2O_3-doped Zirconia (Talk)*. MRS (Boston, USA), 31 November –Ű 3 December 2004.

- B. BUTZ, H. STÖRMER, D. GERTHSEN, A. MÜLLER, and E. IVERS-TIFFÉE. *Correlation between microstructure and degradation of ionic conductivity in Y_2O_3-doped Zirconia (Talk)*. E-MRS (Strasbourg, France), 31 May Ű– 3 June 2005.

- C. PETERS, E. IVERS-TIFFÉE, B. BUTZ, D. GERTHSEN, S. LITZELMAN and H.L. TULLER. *Nanostructured Ionic Materials: Impact on Properties and Performance (Talk)*. E-MRS (Strasbourg, France), 31 May –Ű 3 June 2005.

- B. BUTZ, H. STÖRMER and D. GERTHSEN. *Correlation between Microstructure and Degradation of the Ionic Conductivity in Y_2O_3-doped Zirconia (invited Talk)*. SSI-15 (Baden-Baden, Germany), 17–22 July 2005.

- D. GERTHSEN, E. IVERS-TIFFÉE and H.L. TULLER. *Nanostructured Ionic Materials: Impact on Properties and Performance (Poster)*. SSI-15 (Baden-Baden, Germany), 17–22 July 2005.

- D. GERTHSEN. *D5.1: Electron Microscopy Studies of the Properties of Nanocrystalline Mixed Conductors (Poster)*. Zwischenevaluation CFN (Karlsruhe, Germany), 2005.

- B. BUTZ, C. PETERS, H. STÖRMER, D. GERTHSEN, A. WEBER and E. IVERS-TIFFÉE. *Mikrostrukturelle und elektrische Charakterisierung von nanokristallinen 8.5 mol% Y_2O_3-dotierten ZrO_2 Dünnfilmen (Poster)*. DGM Tag (Karlsruhe, Germany), 21/22 June 2007.

- B. BUTZ, H. STÖRMER and D. GERTHSEN. *Electron Microscopy Studies of Sol-Gel Derived YSZ Thin Films for SOFC Applications (Poster)*. Microscopy Conference MC2007 (Saarbrücken, Germany), 2–7 September 2007.

- B. BUTZ, H. STÖRMER, D. GERTHSEN, C. PETERS, A. WEBER, E. IVERS-TIFFÉE, M. BOCKMEYER and R. KRÜGER. *Microstructural investigations of sol-gel derived 8YSZ thin films (Talk)*. Euromat (Nürnberg, Germany), 10–13 September 2007.

- B. BUTZ, H. STÖRMER and D. GERTHSEN. *Electron Microscopy Studies of Sol-Gel Derived 8YSZ Thin Films for SOFC Applications (Poster)*. Bunsen Colloquium (Clausthal-Zellerfeld, Germany), 27/28 September 2007.

- B. BUTZ, H. STÖRMER, D. GERTHSEN, C. PETERS, A. WEBER, E. IVERS-TIFFÉE, M. BOCKMEYER and R. KRÜGER. *Microstructural and electrical investigations of sol-gel derived 8YSZ thin films (Talk)*. Bunsen Colloquium (Clausthal-Zellerfeld, Germany), 27/28 September 2007.

- S. ROY, B. BUTZ and A. WANNER. *Damage Evolution and Anisotropy in Freeze Cast Metall/Ceramic Composites: an IN-SITU SEM Analysis (Talk)*. 13th European Conference on Composite Materials (ECCM-13) (Stockholm, Sweden), 2–5 June 2008.

- B. BUTZ, H. STÖRMER, D. GERTHSEN, M. BOCKMEYER, R. KRÜGER, P. LÖBMANN, C. PETERS, A. WEBER and E. IVERS-TIFFÉE, . *Microstructure and conductivity of 8.5YSZ thin films obtained by sol-gel processing (Poster)*. Gordon Research Conference GRC: Issues in Grain Boundary Transport: Role of Boundary Chemistry & Structure (Andover, NH, USA), 10–15 August 2008.

- M. DRIES, K. SCHULTHEISS, B. GAMM, B. BUTZ, D. GERTHSEN, B. BARTON and R.R. SCHRÖDER. *Application of a Hilbert phase plate in transmission electron microscopy of materials science samples (Poster)*. European Microscopy Congress (Aachen, Germany), 1–5 September 2008.

- B. BUTZ, R. SCHNEIDER, D. GERTHSEN, M. SCHOWALTER and A. ROSENAUER. *Metastable phase transitions and the degradation of ionic conductivity in the system Y_2O_3-ZrO_2 (Talk)*. SSI-17 (Toronto, Canada), 28 June –Ű 3 July 2009.

- B. BUTZ, R. SCHNEIDER, D. GERTHSEN, M. SCHOWALTER and A. ROSENAUER. *Chemical instability as reason for degradation of ionic conductivity in the system Y_2O_3-ZrO_2 (Talk)*. Microscopy Conference MC2009 (Graz, Austria), 30 August –Ű 4 September 2009

- C. ROCKENHÄUSER, B. BUTZ, N. SCHICHTEL, C. KORTE, J. JANEK and D. GERTHSEN. *Cation transport by dopant (Sm, Gd) diffusion in polycrystalline CeO_2 substrates (Talk)*. Microscopy Conference C2009 (Graz, Austria), 30 August –Ű 4 September 2009.

- L. DIETERLE and B. BUTZ. *Optimized Ar^+-ion etching for TEM cross-section sample preparation (Poster)*. Microscopy Conference MC2009 (Graz, Austria), 30 August –Ű 4 September 2009.

Acknowledgements

Many people have contributed to the success of my work through active collaboration, stimulating discussions or by providing a productive working environment. I thank all of them most sincerely.

First of all, I would like to express my gratitude to Prof. Dr. Dagmar Gerthsen for giving me the opportunity to work at the Laboratory for Electron Microscopy (LEM), Karlsruhe Institute of Technology (KIT). The excellent conditions at her laboratory have made fruitful research possible. Thank you for continuing support and numerous opportunities to attend conferences throughout the last years.

Prof. Dr.-Ing. Ellen Ivers-Tiffée is given thanks for her valuable ideas and for being the co-examiner of this thesis. Furthermore, I greatly thank you for the opportunities to use experimental equipment at the Institute of Materials for Electrical Engineering (IWE), Karlsruhe Institute of Technology (KIT).

I am very grateful to Dr. Reinhard Schneider. His scientific expertise and experimental experience initiated my interest in analytical techniques. He introduced me to the operation of the available microscopes and spectrometers. Our discussions about the physical and experimental backgrounds contributed to a high degree to my understanding and experience in data acquisition and analysis. Thank you for taking the time to support me in Bremen.

I thank Prof. Dr. Andreas Rosenauer for giving me the opportunity to use the recently installed transmission electron microscope at his laboratory at the Institute for Solid State Physics, University Bremen. He and his group, especially Dr. Marco Schowalter and Katharina Gries, made the stays in Bremen a true pleasure for me. Thank you for the nice time.

This work would not have been possible without the close cooperation, in which this work was embedded:

- This work was funded by the Deutsche Forschungsgemeinschaft DFG partly within the common NSF-DFG project *Nanoionics*.

- I thank Christoph Peters, André Weber, and all other members of the group of Prof. Dr.-Ing. Ivers-Tiffée for cooperation in research and fruitful discussions.

- Special thanks appertains the group of the Fraunhofer Institut für Silicatforschung (ISC), Würzburg. Thank you, Dr. Matthias Bockmeyer and Dr.

Reinhard Krüger, for the processing of the electrolyte thin films.

- I thank Dr. Matti Oron-Carl (Institut füür Nanotechnologie (INT), KIT) for the time she spent for the Raman measurements.

- Thanks to Dr. Martina Luysberg (Ernst-Ruska-Centrum für Mikroskopie und Spektroskopie mit Elektronen (ER-C), Forschungszentrum Jülich) for the time she invested in our measurements.

- Many thanks to Scott Litzelman and Prof. Harry Tuller (Crystal Physics and Electroceramics Laboratory, Massachusetts Institute of Technology, Cambridge, United States) for interesting ideas and fruitful discussions.

I would like to thank all of my colleagues at the LEM for the friendly atmosphere, for professional support and discussions, and for pleasant BBQ happenings. In particular, Heike Störmer, Erich Müller, Winfried Send, David Bach, Levin Dieterle, and last but not least Rita Sauter are mentioned here.

I further thank Marina Schramm for the proofreading of this manuscript.

Finally, I would like to thank my friends, my family and in particular my wife Inge Breichler. I am grateful for the pure existence of our daughter Dorothea — this is the privilege of a baby.

Benjamin Butz

Die VDM Verlagsservicegesellschaft sucht für wissenschaftliche Verlage abgeschlossene und herausragende

Dissertationen, Habilitationen, Diplomarbeiten, Master Theses, Magisterarbeiten usw.

für die kostenlose Publikation als Fachbuch.

Sie verfügen über eine Arbeit, die hohen inhaltlichen und formalen Ansprüchen genügt, und haben Interesse an einer honorarvergüteten Publikation?

Dann senden Sie bitte erste Informationen über sich und Ihre Arbeit per Email an *info@vdm-vsg.de*.

Sie erhalten kurzfristig unser Feedback!

VDM Verlagsservicegesellschaft mbH
Dudweiler Landstr. 99
D - 66123 Saarbrücken

Telefon +49 681 3720 174
Fax   +49 681 3720 1749

www.vdm-vsg.de

Die VDM Verlagsservicegesellschaft mbH vertritt

Printed by Books on Demand GmbH, Norderstedt / Germany